Techniques in Differentiation

The discovery of calculus in the seventeenth century by Isaac Newton and Gottfried Leibniz, helped usher in a revolution in mathematics and science that had a profound and far-reaching effect on the world. Calculus provided a powerful tool that enabled the fledgling science of physics to break new ground in our understanding of the workings of the natural universe. Indeed, calculus is virtually synonymous with physics as it is the mathematics of infinitesimal change. As the world about us appears to be a continuity punctuated by discrete things, then calculus is vital in understanding the behaviour of a quantitative change relative to another, from one instant to the next.

The intellectual endeavour of mathematics can be thought of as a tree, with calculus one of its boughs. This bough consisting of two major branches, one entwined about the other—differentiation and integration. This book focuses on the discovery, methods and applications of the mathematics of differentiation. Differential calculus, as opposed to integral calculus, considers variable quantitative relationships to one another in the form of tangents. Integration, on the other hand, is the inverse process that relates quantities as areas and will be explored in the companion volume to this book.

The dedicated student will find in calculus a powerful analytical tool with applications in the physical sciences, engineering and technology. And like all areas of mathematics, it can also be appreciated for its own inherent beauty.

TECHNIQUES IN

DIFFERENTIATION

An Introduction to Elementary Calculus

Volume 1

by S.T. Simms

Perth Academy of Science

Copyright © 2015 by S.T. Simms
Published by Perth Academy of Science
ISBN: 0987528912
ISBN-13: 978-0-9875289-1-9
First edition 2015

Cover design by Pixel Studio
Illustrations by S.T. Simms

To contact publisher: Perth Academy of Science, Suite 27 / 168 Guildford Road, Maylands WA 6051, Australia.
www.mpctutor.com.au

I dedicate this book to my beloved parents,

who gave so much of themselves in loving

and providing for us, their children.

To my siblings and their families who

have shared in this love.

And to my loving wife Mei and our

daughter Jia-Suen—may your life be

rewarded by the pleasure of learning.

ACKNOWLEDGEMENTS

This book has its foundation in an earlier unpublished manuscript with the proposed title, 'Introductory Calculus' and was based upon notes and exercises written for school calculus students at a private education centre in Perth, Western Australia. The author along with Dr A. M. Wilson and Mr M. Handley, all made contributions to those early educational notes.

Table of Contents

Preface

This work has its origins in an earlier manuscript provisionally titled, 'Introductory Calculus', it being based on tutorial notes written for high school calculus students. One motivation for writing this book is partly to fulfil a sense of completeness in finishing the earlier project started by the author many years ago. However, mainly it is the desire to give calculus students a deeper understanding of the subject. Hopefully this is achieved by, in part, providing more historical background and development than is offered by most calculus textbooks.

A common failing of many technical textbooks is to skim over mathematical workings that get to some result. Mathematical and scientific textbooks typically assume the student has the required mathematical skill to provide the missing details for themselves. This is an ongoing major complaint of students and can make the study of a mathematics textbook particularly frustrating. In this book I have endeavoured to provide detailed line-by-line working in proofs and examples. For some it will seem trivial and perhaps an insult to their mathematical ability. To these readers I ask for forbearance and to remember their less capable fellow travellers who may struggle with algebraic technicalities. Of the less capable students some may well find that some of the proofs and examples are still not providing enough detail. To those I apologise and can only suggest they strengthen their foundation in algebra.

Another common complaint of mathematics students is textbooks that provide too few exercises, or overly simple questions with which to practice. I have endeavoured to provide a large number of exercise questions, ranging in level of difficulty from easy to challenging. In addition, this textbook includes the answers to all the questions in the exercises at the end of each chapter. It is particularly irksome when a textbook does not provide answers to exercises—students find it frustrating when they are unable to see if they have adequately mastered the concepts and techniques outlined in a mathematics book.

Finally, very little reliance on technology is assumed for the student in this book. There are some instances when a computer or graphics calculator will be useful, and a scientific calculator is vital for many numerically based questions. But, on the whole, the student should attempt as much as they can with pencil and paper work and only used a calculator when it is absolutely necessary. Therefore, the student can expect answers to questions in this book that require numerical results, given mostly in exact form rather than decimal approximations. It should also be noted that answers to questions are to be fully simplified and so students should assume that arithmetic and algebraic simplification is always required.

Chapter 1

Definition of the Derivative

1.1 The Origin of Differentiation

The mathematical procedure, now known as differentiation, was discovered by Isaac *Newton* (1642–1727) and Gottfried Wilhelm *Leibniz* (1646–1727) during the seventeenth century. The Englishman Newton (Fig. 1.2), was a mathematician and physicist, and is arguably history's greatest scientist. Leibniz (Fig. 1.3), was a philosopher and mathematician from Germany. Their discoveries were based on the work of Pierre de *Fermat* (1608–1665) (Fig. 1.1), a French lawyer in the parliament of Toulouse. Differentiation is a mathematical method by which mathematicians, scientists and engineers can investigate how one quantity can vary relative to another. For example, how the velocity of a moon varies with its distance from a planet or how the acceleration of a rocket changes from moment to moment in time.

In the 1630s and 40s, Fermat and French mathematician Rene *Descartes* independently invented the new mathematical method of co-ordinate geometry, that enabled mathematicians to investigate the geometrical properties of curves using algebra. Utilising co-ordinate geometry and his method of maxima and minima, Fermat tackled two famous problems whose solutions had al-

- 1 -

Fig. 1.1. Pierre de Fermat.

ready been worked out by the ancient Greeks using their method of similar triangles.

These two problems were—finding a *tangent* to a parabola and the problem of maximum area of a rectangular figure of variable sides. Fermat discovered that these two Greek problems could both be solved using the same method. He realised that if he could find a way of analytically determining tangents to a parabola, then he could also solve the problem of finding the maximum area of a variable rectangle. He could achieve this by finding the maximum point of the parabolic curve. Fermat's process involved taking two points very close to each other on a curve and finding the gradient of a connecting line.

The Greeks knew that if you slice a cone with a plane the perimeter of the resulting cross-section will be one of three different curves—the *ellipse*, *parabola* or *hyperbola*, as shown in Fig. 1.4. Now, consider the parabola as shown in Fig. 1.5. Take the point *A* on the axis outside the curve. How do we pick a point *C* on the parabola so that a line, passing through *C* is a tangent to

Fig. 1.2. Isaac Newton.

the parabola? The Greek mathematician and astronomer, *Apollonius* of Perga (c. 262–190 BC), known as the 'The Great Geometer', found that for a parabola the line AE will be tangent when $AO = OB$. Fermat was interested to find if his new mathematical method of coordinate geometry was capable of obtaining this result.

The second Greek problem Fermat was interested in was 'the problem of maximum area'. We can illustrate the problem as

Fig. 1.3. Gottlieb Leibniz.

follows. Suppose we have a line AB of unit length and divide it internally at point C. The line now has two parts AC and BC. The Greeks wanted to know how should point C be chosen so that $AC \cdot BC$ = maximum? The problem is illustrated in Fig. 1.6. Since a product of lengths can be thought of as an *area*, the problem becomes: 'How do you choose point C so that the rectangle having sides AC and BC has the *maximum area*? Through experimentation the Greek mathematicians discovered that the largest area is found when the line AB is divided in half. This result was then proved analytically using their methods of similar triangles.

Fermat discovered that the two Greek problems of maximum area and finding the tangent to a parabola could both be solved using the same method. His first step was to plot a curve showing how area $AC \cdot BC$ varies with the position of point C, with the resulting curve being a *parabola*. This can be shown numerically. Assign $AB = 1$ and $AC = x$. This being the case then $BC = AB - AC = 1 - x$, so that

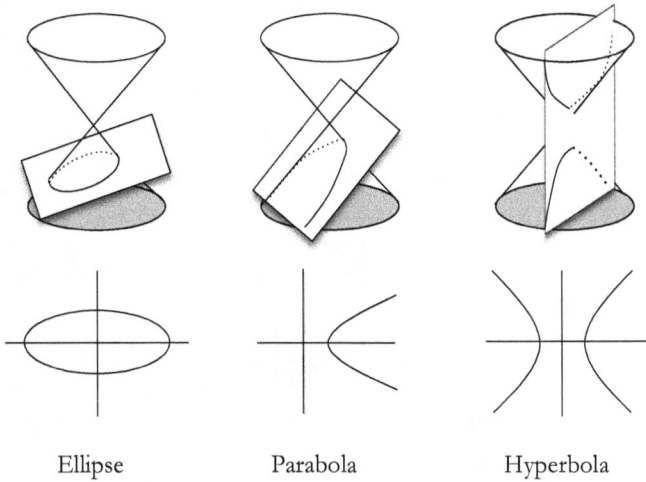

Ellipse   Parabola   Hyperbola

Fig. 1.4. Conic sections.

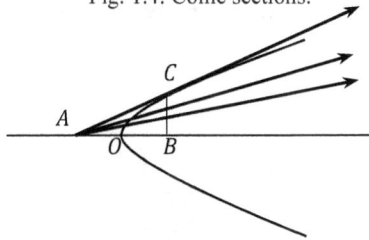

Fig. 1.5. Showing the Greek problem of a tangent to a parabola.

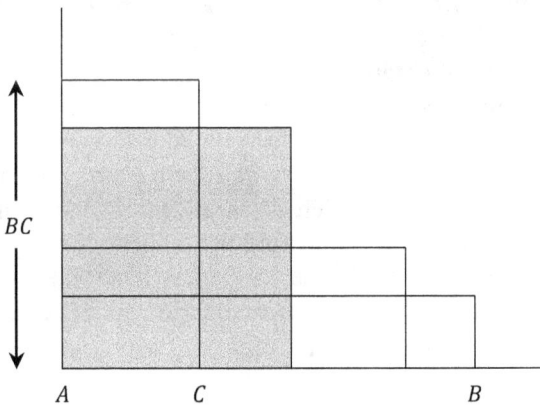

Fig. 1.6. Finding the maximum area.

$$\text{area} = AC \cdot BC = x(1 - x)$$

Now we tabulate the variation of the area with x as shown in Table 1.1. Plotting area against x describes a curve as shown in Fig. 1.7. We see that this curve has a *maximum* at $x = 0.5$ and does in fact describe a *parabola*.

x	$1 - x$	Area $x(1 - x)$
0	1	0
0.1	0.9	0.09
0.2	0.8	0.16
0.3	0.7	0.21
0.4	0.6	0.24
0.5	0.5	0.25
0.6	0.4	0.24
0.7	0.3	0.21
0.8	0.2	0.16
0.9	0.1	0.09
1	0	0

Table 1.1. Showing area values according to x.

Returning to the problem of tangents let us see how this problem and that of the maximum is related. Graphing a tangent to the parabola in Fig. 1.7, we see as x moves away from the point where $x = 0$, the gradient of the tangent decreases until finally at the maximum point where $x = 0.5$ the tangent is horizontal. Fermat therefore realised that if he could find a way of graphing tangents to the parabola, he could also solve the problem of finding the value of x for which the area $x(1 - x)$ is a maximum. All that was required was to find the tangent at each point and select the point at which the tangent is horizontal. This point would be the maximum point of the parabola. Let us explore Fermat's ideas in more detail.

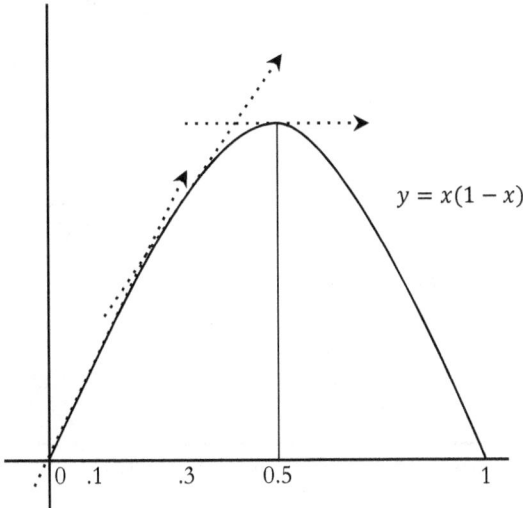

Fig. 1.7. Tangents to the parabolic curve of $y = x(1-x)$.

1.2 Fermat's Method of Maxima and Minima

In order to find the maximum point of a curve Fermat asked himself the following question: *"How does the form of the curve close to a maximum differ from the form of the curve around any other point?"* To find out let us investigate the form of the curve

$$y = x(1 - x)$$

as we zoom in closer to the maximum point at $x = 0.5$ and to another point, say $x = 0.3$. (See Fig. 1.8). We see immediately that as we close in on the maximum point the curve increasingly flattens. However, as we close in to the point at which $x = 0.3$, the curve maintains a non-zero gradient. This observation inspired Fermat to his 'Method of Maxima and Minima'. Fermat thought about it as follows:

We know if we look at an increasingly smaller region around a maximum point the curve becomes flatter. *Finally the curve must become completely flat.* The situation is illustrated in Fig. 1.9. If we move a small distance, h, on either side of the maximum point for $x = 0.5$, then the *value of y essentially does not change at all*. The question then becomes, 'How does this help us find the position of the maximum point?'

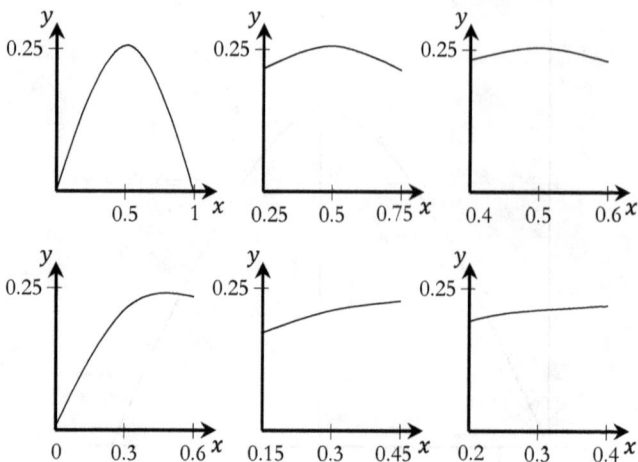

Fig. 1.8. Forms for the curve $y = x(1 - x)$.

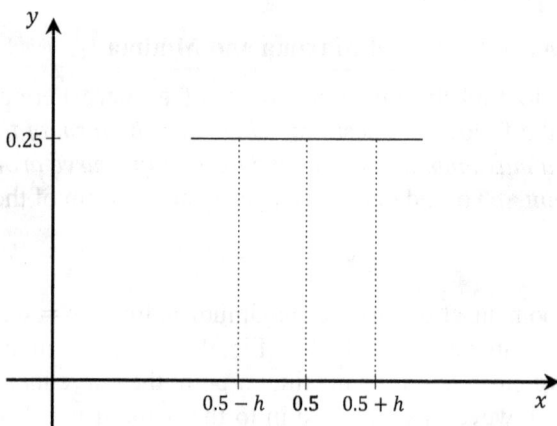

Fig. 1.9. Form of the curve $y = x(1 - x)$ becomes flat
around the maximum point.

Suppose that the maximum point for the curve occurs where $x = a$. That is, the maximum value of y of the curve is given by the function $f(a) = a(1 - a)$. Now for $x = a + h$ the curve has the value $f(a + h) = [a + h][1 - (a + h)]$. Now $f(a)$ is virtually equal to $f(a + h)$ and so their difference is as close to zero as we can imagine. So,

$$f(a + h) - f(a) = 0$$
$$[a + h][1 - (a + h)] - a(1 - a) = 0$$
$$a + h - (a + h)^2 - a(1 - a) = 0$$
$$a + h - a^2 - 2ah - h^2 - a + a^2 = 0$$
$$h - 2ah - h^2 = 0$$

Now dividing by h gives

$$1 - 2a - h = 0$$

Since we made h virtually zero then

$$1 - 2a = 0$$
$$a = 0.5$$

So we have found $x = a = 0.5$ as before.

The minimum of a parabola can also be located using Fermat's method. His method still applies since whether a maximum or minimum, the curve flattens out just the same. Consider the following example.

Example 1.1. *For what value of x will the function* $f(x) = 2x^2 - 8x - 3$ *have a minimum?*

The quadratic function has a minimum since the coefficient[*] of the squared term is positive. We have

$$f(x + h) = 2(x + h)^2 - 8(x + h) - 3$$
$$= 2x^2 + 4hx + 2h^2 - 8x - 8h - 3$$

So,

---

[*] A pre-multiplying factor of a term.

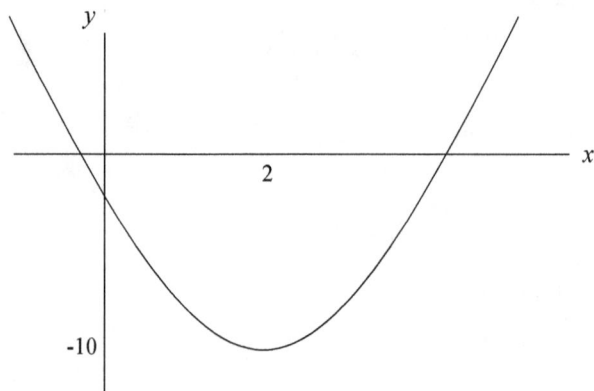

Fig. 1.10. Showing the graph of a minimum turning point parabola.

$$f(x + h) - f(x) = 2x^2 + 4hx + 2h^2 - 8x - 8h - 3$$
$$-(2x^2 - 8x - 3)$$
$$= 4hx + 2h^2 - 8h$$

Equating to zero gives

$$4hx + 2h^2 - 8h = 0$$
$$4x + 2h - 8 = 0$$

Allowing $h = 0$ gives

$$4x - 8 = 0$$
$$x = 2$$

i.e. given function has a minimum at $x = 2$. (See Fig. 1.10).

$\square$

1.3 Fermat's Method of Tangents

We will now consider the behaviour of the parabolic curve at locations other than the maximum or minimum turning points. As discussed in the previous section, the curve does not flatten at these other points but instead maintains a definite non-zero gradient as the point in question is approached. Fermat wished to calculate the gradient of the curve at some particular point. He did so using exactly the same technique as was used to find the maximum or minimum.

Selecting a point on the curve at say, $x = a$, we calculate the gradient of the curve as we focus in on this point. To achieve this we consider a neighbouring point on the curve having x-coordinate $x = a + h$. For the parabola,

$$f(x) = x(1 - x)$$

the y coordinates of the two points are given by

$$f(a) = a(1 - a)$$

and

$$f(a + h) = (a + h)[1 - (a + h)]$$

We can now find the gradient of the chord connecting the two neighbouring points using trigonometry.

From Fig. 1.11 we see that

$$\tan \theta = \frac{BC}{AC}$$

Now,

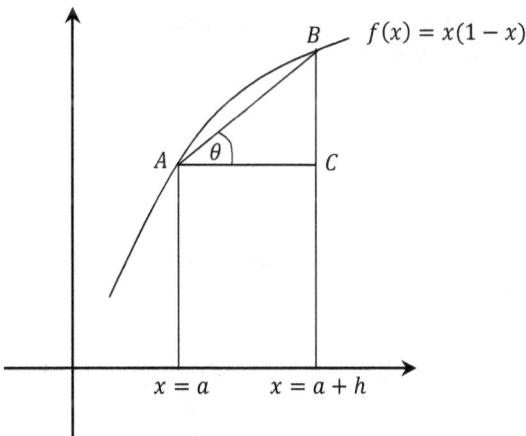

Fig. 1.11. Finding the gradient of a curve.

$$BC = f(a + h) - f(a)$$
$$= (a + h)(1 - a - h) - a(1 - a)$$
$$= a - a^2 - ah + h - ah - h^2 - a + a^2$$
$$= h - 2ah - h^2$$

So,

$$\tan \theta = \frac{h - 2ah - h^2}{h} = 1 - 2a - h$$

We know that the tangent to a curve at a point is the line that just touches but does not *cut* the curve at this point. If the chord here just touches the curve it must have the same gradient as the curve. Fermat now suggested that the non-zero gradient resulting from closing in on point $x = a$ is given by the value of $\tan \theta$ when we set $h = 0$. That is, for our parabolic curve,

$$\tan \theta = 1 - 2a$$

Let us see if Fermat was correct in his assumption by comparing his result to the Greek mathematicians.

To do so we find where the tangent line at $x = a$ cuts the *line of symmetry* of the parabola. The situation is shown in Fig. 1.12. Suppose the tangent that passes through point A intersects the line of symmetry OB at point C. This gives

$$\tan \angle CAB = \frac{BC}{AB}$$

i.e.

$$BC = AB \tan \angle CAB$$

Since

$$\tan \angle CAB = 1 - 2a \quad \text{and} \quad AB = \left(\frac{1}{2} - a\right) = \frac{1}{2}(1 - 2a)$$

we have,

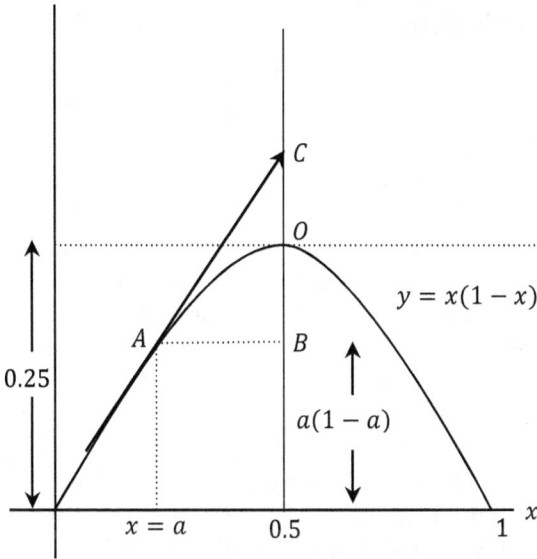

Fig. 1.12. Finding gradient of a tangent.

$$BC = \tfrac{1}{2}(1 - 2a)^2 = \tfrac{1}{2} - 2a + 2a^2$$

Now,

$$OB = \frac{1}{4} - a(1 - a) = \frac{1}{4} - a + a^2$$

That is,

$$OB = \tfrac{1}{2}BC$$

and hence point O is bisecting BC. We therefore see that

$$OB = OC$$

as discovered by Apollonius! Fermat has been proved right; the gradient of the tangent to the curve $y = x(1 - x)$ at point $x = a$ is given by

$$\tan\theta = 1 - 2a$$

1.4 Newton's Discovery of the Differential Calculus

In 1665 a great plague struck the city of London. By 1666 this plague had spread northward to the university town of Cambridge. In the face of this horrific epidemic the university was forced to disband for a time. One of its students, Isaac Newton, then aged 24, returned to his family's farm in Lincolnshire.

Newton had been studying the co-ordinate geometry invented in France 30 years earlier by Descartes and Fermat. He was also acquainted with Fermat's methods of maxima and minima. In the summer of 1666 Newton discovered that Fermat's methods could be extended to *any curve*, not just a parabola. He also saw something even more important.

Initially, Newton simply extended Fermat's method of tangents to more complicated curves. Newton next made an apparently small step that turned out to be momentous for the intellectual fields of both mathematics and physics. Whereas Fermat had focused on the gradient of the curve at a given point, Newton realised by applying Fermat's method of tangents to every point on a curve and plotting the associated gradient, a second curve arises—the *derived curve*, or *derivative*, of the given curve. He wrote fifty years later, "I took Fermat's method and made it more general". Newton called his procedure for obtaining the derived curve the 'Method of Fluxions'. We now call it, the *differential calculus*.

Newton's motivation for developing calculus was to investigate the motion of planetary bodies in their orbits. Building on the work of the German astronomer-scientist *Johannes Kepler*, Newton went on to develop his famous *Law of Gravitation*—a profound scientific synthesis of the force that causes objects to fall to the surface of the earth with the force responsible for the path of astronomical bodies.

Let us apply Newton's method to finding the gradient curve, i.e. the derived curve, for any function $f(x)$. (See Fig. 1.13). Drawing the curve for the function, we place the point A with co-ordinates $(x, f(x))$ on the curve and imagine it to be a fixed point. We now move a small distance h along the horizontal axis, i.e. the x-axis, to $x + h$. Moving vertically upwards the value of the function is $f(x + h)$ giving us point B with co-ordinates

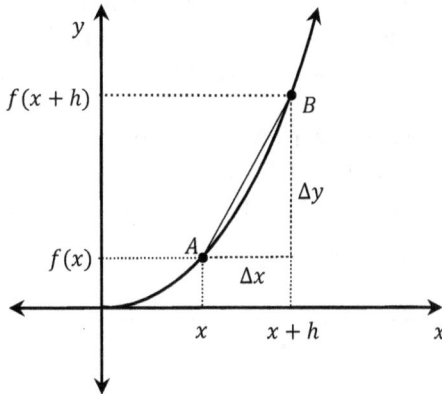

Fig.1.13. Finding the gradient curve.

$(x + h, f(x + h))$. We now find the gradient of the chord AB. The gradient m of a line is defined as the *rise* over the *run*, i.e.

$$m = \frac{rise}{run} = \frac{\Delta y}{\Delta x}$$

with $\Delta x, \Delta y$ representing the *difference* in lengths in the x-direction and the y-direction and thus the length of the sides of the right-angled triangle drawn. Therefore,

$$\frac{\Delta y}{\Delta x} = \frac{f(x + h) - f(x)}{(x + h) - x} = \frac{f(x + h) - f(x)}{h}$$

That is, the gradient of AB is given by the difference in the curve height between points A and B, divided by their horizontal distance apart.

However, our goal is not the gradient of AB but the gradient of the curve itself at any point. So what Newton now did was to take point B, not as a fixed point like A, but as a moveable point, something like a bead on a string, and moved it closer and closer to A. That is, we imagine point B moving infinitely close to A. This makes AB tangent to the curve. In other words, we are finding the *limit* of the ratio $\Delta y/\Delta x$ as Δy and Δx approach zero. We represent this symbolically as

$$\lim_{h \to 0} \frac{f(x + h) - f(x)}{h} \tag{1.1}$$

So, as h goes towards zero the gradient of AB tends toward the gradient of the curve itself. The expression (1.1) therefore describes the gradient of the tangent to a curve for any point on the curve.

Leibniz rediscovered Newton's method around 1674 and introduced the notation dy/dx, where the 'd' represents an infinitely small distance. That is, the derivative, or gradient function, is given by

$$\frac{dy}{dx} = \lim_{h \to 0} \frac{f(x+h) - f(x)}{h} \qquad (1.2)$$

Example 1.2. *Find the derivative of the function* $y = x^2$ *using the Fermat-Newton method.*

Let

$$f(x) = x^2$$

Replacing x with $x + h$ in the function gives us

$$
\begin{aligned}
f(x+h) &= (x+h)^2 \\
&= (x+h)(x+h) \\
&= x^2 + hx + hx + h^2 \\
&= x^2 + 2hx + h^2
\end{aligned}
$$

Substituting into equation (1.2) we find

$$
\begin{aligned}
\frac{dy}{dx} &= \lim_{h \to 0} \frac{x^2 + 2hx + h^2 - x^2}{h} \\
&= \lim_{h \to 0} \frac{2hx + h^2}{h} \\
&= \lim_{h \to 0} (2x + h)
\end{aligned}
$$

Allowing h to equate to 0 we finally obtain

$$\frac{dy}{dx} = 2x$$

□

This new function $2x$ can give us the gradient at any point on the curve described by the function $f(x) = x^2$. That is, it is the gradient function of the original function. Note that this Fermat-Newton method we have just applied is sometimes known as *differentiation by first principles.*

A more complicated curve than the parabola is the *cubic*, where the highest power in the expression describing it is given by x^3. The curve for the simplest representation of this function, $f(x) = x^3$, is shown in Fig. 1.14. Let us find the derivative by first principles of this cubic function.

So,

$$f(x + h) = (x + h)^3$$

$$= (x + h)(x + h)^2$$

$$= (x + h)(x^2 + 2hx + h^2)$$

$$= x^3 + 2hx^2 + h^2x + hx^2 + 2h^2x + h^3$$

$$= x^3 + 3hx^2 + 3h^2x + h^3$$

Therefore,

$$\frac{dy}{dx} = \lim_{h \to 0} \frac{x^3 + 3hx^2 + 3h^2x + h^3 - x^3}{h}$$

$$= \lim_{h \to 0} \frac{3hx^2 + 3h^2x + h^3}{h}$$

$$= \lim_{h \to 0}(3x^2 + 3hx + h^2)$$

$$= 3x^2$$

Example 1.3. *Find the derivative of the function*

$$f(x) = x^3 + 4x^2 - 3x + 5$$

We have

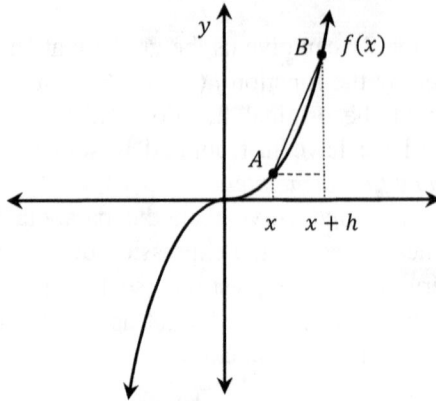

Fig. 1.14. Showing the cubic function.

$$f(x + h) = (x + h)^3 + 4(x + h)^2 - 3(x + h) + 5$$
$$= (x^3 + 3hx^2 + 3h^2x + h^3) + 4(x^2 + 2hx + h^2) - 3(x + h)$$
$$+ 5$$
$$= x^3 + 3hx^2 + 3h^2x + h^3 + 4x^2 + 8hx + 4h^2 - 3x - 3h + 5$$

So,

$$f(x + h) - f(x) = 3hx^2 + 3h^2x + h^3 + 8hx + 4h^2 - 3h$$

We now use a notation introduced by the Italian-French astronomer and mathematician *Joseph-Louis Lagrange* (1736–1813), who represented the derivative of a function as $f'(x)$. We find,

$$f'(x) = \lim_{h \to 0} \frac{3hx^2 + 3h^2x + h^3 + 8hx + 4h^2 - 3h}{h}$$
$$= \lim_{h \to 0}(3x^2 + 3hx + h^2 + 8x + 4h - 3)$$
$$= 3x^2 + 8x - 3$$

□

In order to continue our *differentiation* to higher powers of x, we now make use of the beautiful pattern discovered by Indian mathematicians in the 10th century A.D. We notice that

$$(x + h)^1 = 1x + 1h$$

$$(x + h)^2 = 1x^2 + 2hx + 1h^2$$

$$(x + h)^3 = 1x^3 + 3hx^2 + 3h^{2x} + 1h^3$$

$$(x + h)^4 = 1x^4 + 4x^3h + 6x^2h^2 + 4xh^3 + 1h^4$$

$$(x + h)^5 = 1x^5 + 5x^4h + 10x^3h^2 + 10x^2h^3 + 5xh^4 + 1h^5$$

and so on. Looking at the numbers in the expansions here it can be seen that these coefficients form a pattern. Following this pattern we can therefore obtain the coefficients of the terms in the expansions of $(x + h)^n$, where n represents a counting number. Notice too that the powers on the x decrease by one for each successive term and increase on h by one.

French mathematician *Blaise Pascal* (1623–1658) discovered that that these coefficients have a *combinatorial* meaning. We now call this triangle of numbers *Pascal's triangle*, and is usually written as

$$
\begin{array}{ccccccccccc}
 & & & & & 1 & & & & & \\
 & & & & 1 & & 1 & & & & \\
 & & & 1 & & 2 & & 1 & & & \\
 & & 1 & & 3 & & 3 & & 1 & & \\
 & 1 & & 4 & & 6 & & 4 & & 1 & \\
1 & & 5 & & 10 & & 10 & & 5 & & 1
\end{array}
$$

This pattern can continue ad infinitum, with the numbers in the triangle formed by the sum of the two numbers diagonally above it. Each row forms the numerical coefficients in a binomial[†] expansion. As Pascal noticed, these numbers are also formed with *combinatorials* defined by

---

† Meaning two terms

$$\binom{n}{r} = \frac{n!}{(n-r)!\, r!}$$

where $n!$ (pronounced n factorial) is given by

$$n! = n(n-1)(n-2)(n-3)\dots 1$$

Here, $n!$, tells us the number of ways to order n objects and $\binom{n}{r}$ the number of ways to choose r objects from n objects when order of selection doesn't matter. In Pascal's triangle the row number gives the value of n and the value of r is given by the position number starting from the left; numbering starts at zero. Thus for example, the fourth row down gives coefficients,

$$\binom{3}{0} = \frac{3!}{3!\,0!} = \frac{3.2.1}{3.2.1.1} = \frac{6}{6} = 1$$

$$\binom{3}{1} = \frac{3!}{2!\,1!} = 3$$

$$\binom{3}{2} = \frac{3!}{1!\,2!} = 3$$

$$\binom{3}{3} = \frac{3!}{0!\,3!} = 1$$

‡

Let us make use of Pascal's triangle to find the derivative for the general polynomial function $y = x^n$. We have

$$\binom{n}{0} = \frac{n!}{n!\,0!} = 1$$

$$\binom{n}{1} = \frac{n!}{(n-1)!\,1!} = \frac{n(n-1)!}{(n-1)!} = n$$

---

‡ Note that 0!=1 since there is one way to order 0 objects.

$$\binom{n}{2} = \frac{n!}{(n-2)!\,2!} = \frac{n(n-1)(n-2)!}{(n-2)!\,2} = \frac{n(n-1)}{2}$$

$$\vdots$$

$$\binom{n}{n-1} = \frac{n!}{1!\,(n-1)!} = n$$

$$\binom{n}{n} = \frac{n!}{0!\,n!} = 1$$

Now we can more easily expand $(x+h)^n$, i.e.

$$(x+h)^n = \binom{n}{0}x^n h^0 + \binom{n}{1}x^{n-1}h^1 + \binom{n}{2}x^{n-2}h^2 + \cdots$$

$$+ \binom{n}{n-1}x^1 h^{n-1} + \binom{n}{n}x^0 h^n$$

$$= x^n + nx^{n-1}h + \frac{n(n-1)}{2}x^{n-2}h^2 + \cdots + nxh^{n-1}$$

$$+ h^n$$

Hence,

$$\frac{(x+h)^n - x^n}{h} = nx^{n-1} + \frac{n(n-1)}{2}x^{n-1}h + \cdots + nxh^{n-2}$$

$$+ h^{n-1}$$

Taking the limit as $h \to 0$ gives us a formula for the derivative of expressions of the form

$$y = x^n$$

i.e.

$$\boxed{\frac{dy}{dx} = nx^{n-1}}$$

This *power rule* provides the key to the efficient use of differential calculus.

Example 1.4. *Using the 'shortcut' power rule formula find the derivative of* $y = x^7$:

$$\frac{dy}{dx} = 7x^{7-1} = 7x^6$$

$\square$

Notice in deriving the power rule above that if the function in question had a coefficient other than 1, that number could simply be factorised out of the expansion. So, finding the derivative for functions of the form $y = ax^n$, where a represents a constant, the derivative becomes

$$\boxed{\frac{dy}{dx} = anx^{n-1}}$$

Note what happens when we find the derivative of a constant only. We can think of the constant as being in product with x^0, since of course, $x^0 = 1$. So the derivative for $y = a$, for example, becomes

$$\frac{dy}{dx} = 0 \times ax^{0-1} = 0$$

In addition, it should be noted that finding the derivative of an expression consisting of terms we simply find the derivative of each term in turn. That is, the derivative of a sum or difference is simply the derivative of each term taken individually. The following example will illustrate this.

Example 1.5. *What is the derivative of the function*

$$f(x) = 3x^{10} + x^5 - 7x^3 - 15 ?$$

Solution:

$$f'(x) = 10 \times 3 \times x^{10-1} + 5 \times x^{5-1} - 3 \times 7 \times x^{3-1} -$$
$$= 30x^9 + 5x^4 - 21x^2$$

$\square$

The power rule also works for *negative* powers of n. Replacing $-n$ for n we find the derivative for $y = ax^{-n}$:

$$\frac{dy}{dx} = -anx^{-n-1}$$

e.g. for $y = 1/x^3$, finding the derivative:

$$y = x^{-3}$$

$$\therefore \frac{dy}{dx} = -3x^{-4} = \frac{-3}{x^4}$$

Just as taking the derivative gives us the gradient function, there is no reason why we cannot find the derivative of the gradient function itself to give what is called the *second derivative*.

Example 1.6. *Find the second derivative of the function* $f(x) = -x^2 + 6x + 10$.

We have

$$f(x) = -x^2 + 6x + 10$$
$$f'(x) = -2x + 6$$
$$f''(x) = -2$$

$\square$

For *higher derivatives* the Leibniz notation is written $\frac{d^n y}{dx^n}$ where n is a positive integer. Using this notation let us take the second derivative of the parabolic function $f(x) = x(1 - x)$ [see Fig. 1.15]. We have for the first derivative,

$$\frac{df}{dx} = \frac{d}{dx}[x(1 - x)] = \frac{d}{dx}[x - x^2] = 1 - 2x$$

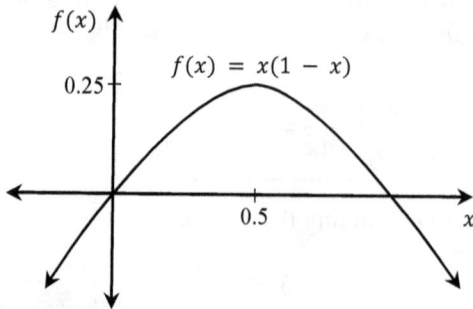

Fig. 1.15. Showing curve of $f(x) = x(1 - x)$.

That is, the first derivative of this *parabola* is the *straight line* $y = 1 - 2x$ (Fig. 1.16). Applying the derivative procedure again we find

$$\frac{d^2f}{dx^2} = \frac{d}{dx}\left(\frac{df}{dx}\right) = \frac{d}{dx}(1 - 2x) = -2$$

The second derivative of the parabolic curve $y = x(1 - x)$ is therefore the line $y = -2$, that is, a line parallel to the x-axis (Fig. 1.17).

We can apply this process to even higher-powered functions and their derivatives, e.g. $f(x) = x^3$. We have

$$f'(x) = 3x^2$$

so that

$$f''(x) = 3\left[\frac{d}{dx}(x^2)\right] = 3 \cdot 2 \cdot x = 6x$$

and

$$f'''(x) = 6\left[\frac{d}{dx}(x)\right] = 6 \cdot 1 = 6$$

with all higher derivatives vanishing. (See Fig. 1.18).

Newton's calculus technique allows us to easily find points on a curve of a particular gradient size. For example, as previ-

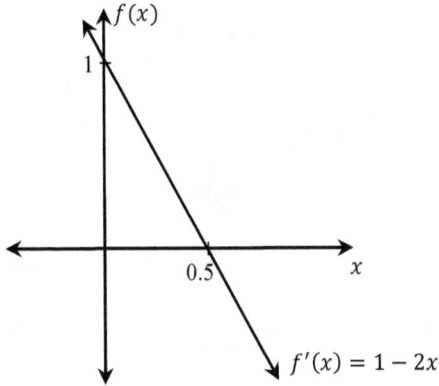

Fig. 1.16. Gradient function of $f(x) = x(1 - x)$.

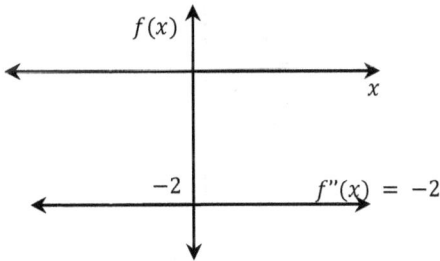

Fig. 1.17. Graph of second gradient function of $f(x) = x(1 - x)$.

ously seen, turning points on a curve have a zero gradient. All that is required to locate a turning point is to find the gradient function by taking the derivative, equating to zero and solving.

Example 1.7. *Use the power rule to find the turning point of the function in example 1.1.*

Differentiating with respect to x we have

$$f'(x) = 4x - 8$$

Equating to zero gives,

$$4x - 8 = 0$$
$$4x = 8$$
$$x = 2$$

Now,

$$f(2) = 2(2)^2 - 8(2) - 3$$
$$= 8 - 16 - 3 = -11$$

That is, a turning point exists at $(2, -11)$.

□

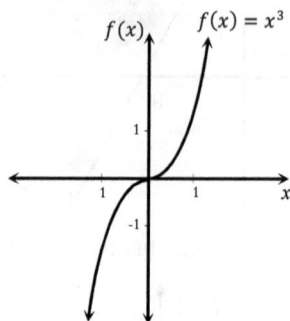

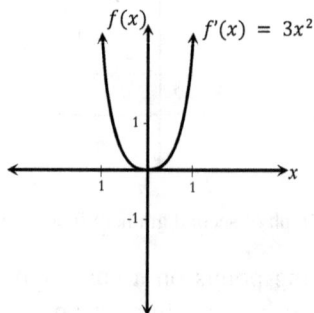

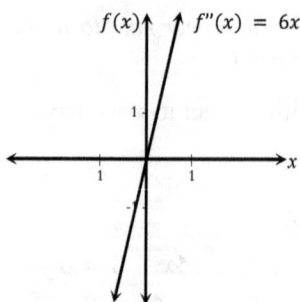

Fig. 1.18. Showing the curves for $f(x) = x^3$ and its derived functions.

Exercise 1

1. Find from *first principles* the derivative of the following expressions:

   (a) $10x^2$ (b) $x^2 - 3x + 5$ (c) x^3 (d) $(3x + 1)^4$

   (e) $2x^3 + 4x + 1$ (f) x^{-n} (g) $\sqrt{x}$ (h) $x^{\frac{3}{2}}$

   (i) $\frac{1}{(1-x)^4}$ (j) $(2 - 5x)^{-\frac{1}{4}}$ (k) $\frac{a-x}{x^2}$

2. Prove that the derivative of a constant is zero.

3. Use the power rule to find the derivative of the following expressions:

   (a) x^7 (b) x^{3n} (c) $7x^3$ (d) $\frac{5}{2}x^{12}$ (e) $10x$ (f) 5

   (g) $(6x)^2$ (h) $x^4 \times x^7$ (i) $\frac{5}{x}$ (j) $7\sqrt{x}$

   (k) $5x^3 + 10x^2$ (l) $4(2x - 1)^2$ (m) $2x - 7$

   (n) $\frac{3x^2 - x}{x^2}$ (o) $3x^3 - 5x^2 + 6x - 4 - \frac{2}{x}$

4. Given a function $f(x) = 2x^3 + 5x^2 - x$, find the values of $f'(0), f'(2), f'(-1), f''(-2), f'''(1)$.

5. Find the value(s) of x for which the derivative for the function

$$f(x) = \frac{x^3 + 6x + 11}{x}$$

   is zero. What is the geometrical interpretation of this result?

6. Find the points on the curve described by

$$y = 2(x - 1)(x + 2)(\tfrac{1}{2}x - 2)$$

   where the tangent of the curve is parallel to the x-axis. Graph the curve and note the features on it where the gradient is zero.

7. The path of an object is described by the equation

$$s = 5t - 4.9t^2$$

Find the object's turning point.

8. If $s = 12t - 5t^2$, find the value of s for which $\frac{ds}{dt} = 0$.

9. How many derivatives are required to reduce the expression $\frac{1}{2}x^5$ to a constant?

Answers

1) a) $20x$ b) $2x - 3$ c) $3x^2$ d) $12(3x + 1)^3$ e) $6x^2 + 4$

f) $-\frac{n}{x^{n+1}}$ g) $\frac{1}{2\sqrt{x}}$ h) $\frac{3\sqrt{x}}{2}$ i) $\frac{4}{(1-x)^5}$ j) $\frac{5}{4(2-5x)^{\frac{5}{4}}}$ k) $\frac{x-2a}{x^3}$

3) a) $7x^6$ b) $3nx^{3n-1}$ c) $21x^2$ d) $30x^{11}$ e) 10 f) 0 g) $72x$

h) $11x^{10}$ i) $-\frac{5}{x^2}$ j) $\frac{7}{2\sqrt{x}}$ k) $15x^2 + 20x$ l) $16(2x - 1)$ m) 2

n) $\frac{1}{x^2}$ o) $9x^2 - 10x + 6 + \frac{2}{x^2}$

4) -1, 43, 4, -14, 12

5) $\sqrt[3]{\frac{11}{2}}$

6) $(1 \pm \sqrt{3}, \pm 6\sqrt{3})$

7) $(0.51, 1.28)$

8) $7\frac{1}{5}$

9) 5

Chapter 2

Simple Applications
of Differentiation

2.1 Rectilinear Motion

Let us now see what happens when we translate from mathematics to physics, as Newton himself would have done. The simplest form of motion in physics is often referred to as rectilinear motion. The word rectilinear comes from the Latin: 'rectus' = straight, 'linea' = line. Rectilinear motion then, literally means motion in a straight line. We can use calculus to investigate the motion of objects, in particular, objects moving in a straight path.

Suppose we let the variable x in the parabolic function $f(x) = x(1 - x)$ from the previous chapter represent *time*. That is, with $x = t$ let $f(x) = s(t)$ represent the *distance* travelled by a particle from a fixed point in time t. Then

$$\frac{df}{dx} \rightarrow \frac{ds}{dt} = 1 - 2t$$

Consider the meaning of the derivative ds/dt. It measures the gradient of the curve $s(t)$ at time t and therefore tells us the *rate at which the distance $s(t)$ is changing with time*. But this rate is simply the speed, or *velocity*. We therefore have

$$\text{velocity } v(t) \text{ at time } t = \frac{ds}{dt} = 1 - 2t$$

The change of quantities relative to time is often referred to in calculus as simply *rates of change*. In particular, in this example we are talking about *instantaneous rates of change*. Before the invention of calculus scientists could only determine the *average rate of change*—taking the difference between the beginning and end values of a variable quantity and dividing by the elapsed time. E.g.

$$\text{average velocity } = \frac{\Delta v}{\Delta t} = \frac{v_1 - v_0}{t_1 - t_0}$$

It turns out that the average of all the instantaneous rates of change of a quantity is simply equal to the average rate of change. Or, in geometrical terms, the average gradient of a curve between two points is simply the gradient of the chord joining these two points.

Next, consider the second derivative

$$\frac{d^2 f}{dx^2} \rightarrow \frac{d^2 s}{dt^2} = -2$$

Consider the meaning of $d^2 s / dt^2$. We know that

$$\frac{d^2 s}{dt^2} = \frac{d}{dt}\left(\frac{ds}{dt}\right) = \frac{d}{dt}(v(t))$$

The derivative $d^2 s / dt^2$ therefore tells us *the rate at which the velocity $v(t)$ is changing with time*. But this rate is simply the *acceleration*. We therefore have

$$\text{acceleration } a(t) \text{ at time } t = \frac{d^2 s}{dt^2} = -2$$

Our distance function

$$s(t) = t(1 - t)$$

therefore describes the motion of a particle which moves with *constant negative acceleration* $a(t) = -2$. This is exactly what happens for an object moving near the surface of the Earth. The force of gravity imparts to such an object a constant downward acceleration of 9.8 metres per second per second, i.e. -9.8 m/s^2, the negative indicating the direction, i.e. downwards.

Example 2.1. *If a bullet is fired from a gun with an initial velocity of 300 metres per second directly upwards, its height at the end of t seconds is described by the equation* $h = 300t - 4.9t^2$. *Ignoring air resistance, how long will it take for the bullet to stop rising?*

The displacement[§] is given by

$$h = 300t - 4.9t^2$$

$$\therefore v = \frac{dh}{dt} = 300 - 2(4.9)t = 300 - 9.8t$$

Obviously the velocity of the bullet will be zero at the instant it stops rising, so

$$v = 300 - 9.8t = 0$$

i.e.

$$t = \frac{-300}{-9.8} = 30.6 \text{ sec}$$

So our bullet takes 30.6 seconds before it slows down to zero velocity. The height at which this occurs is $h = 300(30.6) - 4.9(30.6)2 = 4.6$ km! This is not a realistic situation however, as we have ignored air resistance which would reduce the time of flight and maximum height attained.

$\square$

Example 2.2. *The displacement of a particle along a straight line is given by the function*

$$s = 2t^3 - 8t^2 + 75t + 98$$

[§] i.e. change in position.

What are its velocity and acceleration functions at any time t?

We have

$$v(elocity) = \frac{ds}{dt} = 6t^2 - 16t + 75$$

$$a(cceleration) = \frac{dv}{dt} = \frac{d^2s}{dt^2} = 12t - 16$$

□

Example 2.3. *The motion of a body is described by the equation of motion*

$$x = t^3 - 9t^2 + 15t$$

where x is the body's displacement in metres. What is the body's instantaneous velocity and acceleration at time t = 2 seconds? In addition, find the distance covered by the body between the times it is at zero velocity.

$$v = \frac{dx}{dt} = 3t^2 - 18t + 15$$

and

$$a = \frac{d^2x}{dt^2} = 6t - 18$$

So at $t = 2$,

$$v(2) = 3(2)^2 - 18(2) + 15 = -9 \text{ m/s}$$

$$a(2) = 6(2) - 18 = -6 \text{ m/s}^2$$

Turning to the second part of the question now, let $v = 0$ and solve for t:

$$3t^2 - 18t + 15 = 0$$

$$t^2 - 6t + 5 = 0$$

$$(t - 1)(t - 5) = 0$$

$$\therefore t = 1, 5 \text{ s}$$

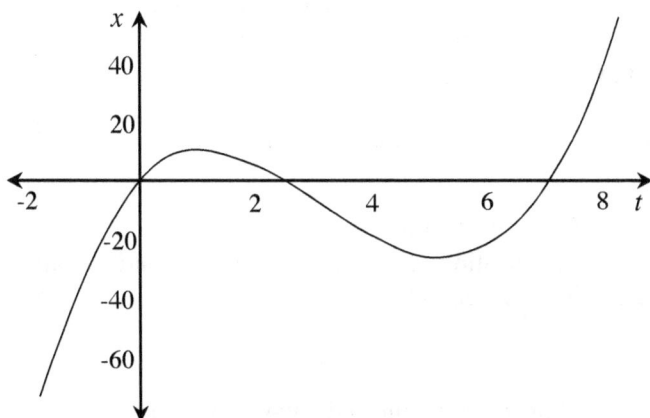

Fig. 2.1. Graph of $x = t^3 - 9t^2 + 15t$.

Notice how the graph of the displacement function, as shown in Fig. 2.1, shows a change in direction. That is, at just after 2 seconds the body begins going in a negative direction. To find the total distance traversed we will need to find the maximum displacements in both the positive and negative directions within the domain $1 < t < 5$. In other words, we now determine the location of the turning points of the curve in a similar manner to Fermat's method in chapter 1.

To find these extrema we determine the gradient function and equate to zero. However, we have already found the gradient function; it is simply the velocity. So, the extrema are located at the times calculated above, namely $t = 1$ and $t = 5$. Thus,

$$x(1) = 1 - 9 + 15 = 7$$

$$x(5) = 5^3 - 9(5)^2 + 15(5) = -25$$

Therefore, from its extremum at times $t = 1$ and 5, the body moves a distance of 7 metres to reach its origin before moving another 25 metres to its other extremum position. That is, it moves a total of 32 metres in the time.

Note that the object in question is actually travelling in one dimension only. So we could consider it moving back and forward along the x-axis. This can be a little difficult to visualise so

in Fig. 2.1 above we have graphed its motion in two dimensions—the x-direction plotted as the vertical axis and time repre-
sented by the horizontal axis.

2.2 Equation of a Tangent to a Curve

In the previous chapter we saw how the derivative derives the
gradient of a tangent to a curve. It is only a short step to then
determine the equation of the tangent. The standard equation for
a straight line is given by

$$y = mx + c$$

m representing the gradient and c the y-intercept. Knowing the
co-ordinates of the point of interest on a curve we obtain m from
the derivative of the curve at this point. Since the tangent to the
curve at this point obviously includes the point within itself, we
can simply determine c by substituting the values of x and y for
the point. For example, say we wish to find the equation of the
tangent to the curve $y = x^3$ at the point $(2, 8)$. (See Fig. 2.2).
We take the derivative to obtain the gradient expression $3x^2$. For
$x = 2$, this expression equals 12. So the equation of the tangent
is

$$y = 12x + c$$

Substituting $(2, 8)$ into this equation enables us to determine c,
i.e. $c = -16$, giving

$$y = 12x - 16$$

Example 2.4. *Find the equation of the tangent to the curve*
$y = -\frac{1}{3}x^2 + \frac{1}{2}x + 3$ *at the point* $(1, 3)$.

Take the derivative with respect to x,

$$\frac{dy}{dx} = -\frac{2}{3}x + \frac{1}{2}$$

So the gradient of the tangent is given by

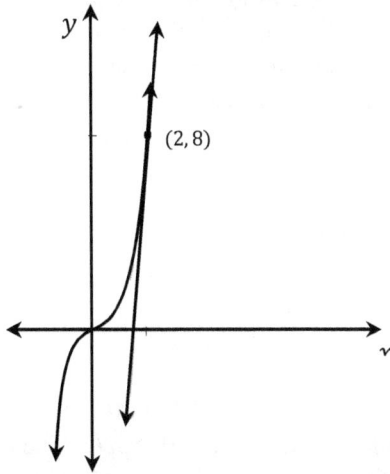

Fig. 2.2. Finding equation of tangent.

$$m = \left[\frac{dy}{dx}\right]_{x=1} = -\frac{2}{3}(1) + \frac{1}{2} = -\frac{1}{6}$$

Thus the equation of the tangent becomes

$$y = -\frac{1}{6}x + c$$

To find the y-intercept of the tangent line substitute $(1, 3)$ giving

$$3 = -\frac{1}{6}(1) + c$$

i.e.

$$c = 3 + \frac{1}{6} = \frac{19}{6}$$

Therefore,

$$y = -\frac{1}{6}x + \frac{19}{6}$$

or,

$$x + 6y = 19$$

□

We can also find the gradient and equation of the *normal* to a curve. The normal is simply a line that is perpendicular to a tangent of a curve. The relationship between the gradients of perpendicular lines can be found in the following manner. Consider the right-angled triangle ABC in Fig. 2.3. We have

$$\tan \theta = \frac{a}{b}$$

and

$$\tan \phi = \frac{b}{a}$$

We know that the gradient of line AC is given by the rise over the run, $m = h/l$, and also $\tan \theta = h/l$. Therefore, the gradient of a line equals to the tan of the angle the line makes with the x-axis. By this we also find that the gradient of BC is given by $\tan \psi$. Now, from the trigonometry of the unit circle we know that $\tan \psi = -\tan \phi$. Therefore, the gradient of BC equals $-\tan \phi$. So, since

$$m_{AC} = \tan \theta = \frac{a}{b} \quad \text{and} \quad m_{BC} = -\tan \phi = -\frac{b}{a}$$

then

$$m_{AC} = -\frac{1}{m_{BC}}$$

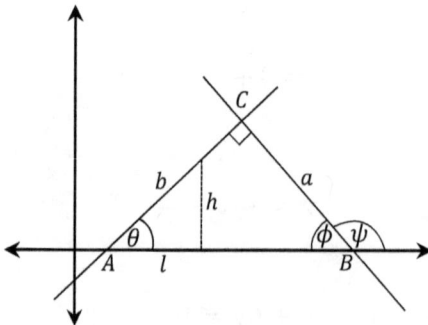

Fig. 2.3. Determining gradient of perpendicular line.

We can summarise by saying the gradients of perpendicular lines are the negative reciprocal of each other. i.e. for perpendicular lines 1 and 2:

$$m_1 = -\frac{1}{m_2}$$

or,

$$m_1 m_2 = -1$$

2.3 Increments and Differentials

In the previous chapter we saw that the derivative fraction dy/dx was derived from the gradient fraction of a chord to a curve, $\Delta y/\Delta x$, by simply making the differential Δx approach zero. From Fig. 2.4 we see that the gradient of the chord PQ is given by $\delta y/\delta x$, where $\delta x, \delta y$ are increments in the x and y directions. For small values of $\delta x, \delta y$ the gradient of the chord is approximately equal to the gradient of the tangent. That is,

$$\frac{\delta y}{\delta x} \approx \frac{dy}{dx}$$

so,

$$\delta y \approx \frac{dy}{dx} \cdot \delta x$$

Thus, we can use this relationship to estimate *small changes* in a quantity based on the small change in another related quantity, and knowing the derived function.

Example 2.5. *The radius of a sphere is 5 cm. What is the increase in volume of the sphere if its radius is increased by 1 mm?*

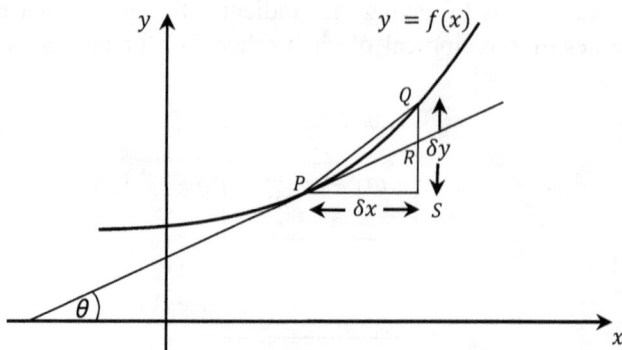

Fig. 2.4. Showing the relationship between increments and differentials.

The volume of a sphere is given by the formula,

$$V = \frac{4}{3}\pi r^2$$

So,

$$\frac{dV}{dr} = 4\pi r^2$$

Since,

$$\delta V \approx \frac{dV}{dr} \cdot \delta r$$

then

$$\delta V \approx 4\pi r^2 \cdot \delta r$$

Given $r = 5$ cm and the small change in r of 0.1 cm, then

$$\delta V \approx 4\pi(5)^2 \times 0.1 = 10\pi$$

That is, the volume of the sphere increases in size by approximately 10π, or 31.4 cm³.

□

Example 2.6. *Find the approximate percentage error in the determination of the surface area of a ball bearing of diameter 10 mm, if the measurement of its radius was in error by 0.1 mm.*

Recalling that the area of a sphere is given by $A = 4\pi r^2$ we find its approximate change is determined using

$$\delta A \approx \frac{dA}{dr} \cdot \delta r = 8\pi r \delta r$$

Now dividing each side by A and multiplying by 100 we find,

$$\frac{\delta A}{A} \times 100 = \frac{800\pi r \delta r}{A}$$

$$= \frac{800\pi r \delta r}{4\pi r^2}$$

$$= \frac{200\delta r}{r}$$

The left hand side represents the percentage error in the area. Thus substituting $r = 5$ and $\delta r = 0.1$ we obtain

$$\text{percentage error} = \frac{200 \times 0.1}{5} = 4\%$$

□

We make two additional observations of Fig. 2.4: As $x \to x + \delta x$, then

$$f(x) \to f(x) + \delta y$$

i.e.

$$f(x) \to f(x) + \frac{dy}{dx} \cdot \delta x$$

Secondly, notice from triangle PRS we have,

$$\tan\angle RPS = \frac{SR}{SP}$$

Since $\angle RPS = \theta$ and the ratio SR/SP equals the gradient of PR, then the gradient of the tangent line equals $\tan\theta$. That is, $\frac{dy}{dx} = \tan\theta$. So, knowing the angle of intersection a line makes with the x-axis, the equation of the line can be determined using

$$y = \tan\theta \cdot x + c$$

where c is the value of the intercept of the line with the y-axis.

Example 2.7. *Find the angles at which the tangents to the cubic curve $y = \frac{1}{3}x(x+2)(x-3)$ at the x-axis, make with the x-axis.*

It will be simpler to take the derivative by first expanding the brackets in the cubic function. So,

$$y = \frac{1}{3}x(x+2)(x-3) = \frac{1}{3}x^3 - \frac{1}{3}x^2 - 2x$$

Differentiating gives,

$$\frac{dy}{dx} = x^2 - \frac{2}{3}x - 2$$

Now, the curve intersects the x-axis where $y = 0$, i.e.

$$\frac{1}{3}x(x+2)(x-3) = 0$$

to give $x = 0, -2,$ or 3. The gradient of the tangents of the curve at these points is given by,

$$\left.\frac{dy}{dx}\right|_{x=0} = -2$$

$$\left.\frac{dy}{dx}\right|_{x=-2} = (-2)^2 - \frac{2}{3}(-2) - 2 = \frac{10}{3}$$

$$\left.\frac{dy}{dx}\right|_{x=3} = (3)^2 - \frac{2}{3}(3) - 2 = 5$$

From $\tan\theta = -2, \frac{10}{3}, 5$ we obtain $\theta = 116.6°, 73.3°, 78.7°$. That is, moving left to right along the x-axis we find the cubic curve intersecting the x-axis at angles $73.3°, 116.6°$ and $78.7°$ respectively.

□

Exercise 2

1. Find the equation of the tangent to the curve $y = -3x^2 + 5$ at the point $(1, 2)$.

2. Find the equation of the line parallel to $y = 3x - 4$ and passing through the point $(-5, 2)$.

3. Find the equation of the line *normal* to $y = 3x + 4$ and containing $(2, -1)$.

4. Find the equation of the tangent and its normal to the parabola $y = 5x^2 + 2$ at the point $(-1, 3)$. Confirm that the lines intersect at this same point by performing simultaneous equations.

5. Find the equation of the tangent and a line perpendicular to it at the point $(4, 3)$ on the cubic curve $y = \frac{1}{2}x^3 - 2x$.

6. Plot the curve $y = \frac{(x-2)^3}{5}$ over the domain $-1 \le x \le 5$. Determine the gradient and graph the tangent at the point where $x = 1$.

7. Sketch the curve $3y = x^2 - 10$ and its tangent at the point $(5, 5)$, and find the area of the triangle formed from the tangent's axis intercepts and the origin.

8. Find the co-ordinates of the intersection points of the curves $y = 3x^2$ and $y = x^2 - 2x + 3$. What are the equations of the tangents at these points?

9. The curve $y = ax^2 + bx + c$ passes through $(-1, 9)$ with a gradient of -5 at that point. Find the values of the coefficients given a y-intercept of 3.

10. Consider the relation for an ellipse $x^2 - 2x + 4y^2 = 0$. Find the equation of the line running tangent to the described ellipse at the point $(2, \frac{1}{2})$.

11. Consider the parabola $y = 2x^2$. For $x = b$ and $x = c$, $b < 0$, $c > 0$, find the relationship between b and c for tangents to the curve that are right angles to each other.

12. Find the intersection point of the lines that touch the curve $y = 2x^3 - 6x^2 + 5x + 10$ at the points $(-2, -40)$ and $(3, 25)$.

13. Find the equations of the tangent lines to the curve $y = 2x - 4x^3 + 6$ that lie normal to the line $2x + 3y = 1$.

14. Find the equation of the tangent passing through the point $(1, 1)$ on the curve $16y = x^3 - 4x^2 - 11x + 30$. Also find the other intersection points of this tangent with the curve.

15. The function $f(x) = ax^2 + bx + c$ describes the path of a projectile moving near the Earth's surface. At $(1, 4)$ the tangent to its path is described by $y - 3x = 1$. The projectile also passes through $(10, 3)$. Use this information to find the value of the constants a, b, c.

16. Show that the curve $y = \frac{1}{4}x^4 - \frac{7}{3}x^3 + 6x^2$ has three tangents parallel to the x-axis.

17. What is the initial displacement of a particle that has its motion described by

$$x = 3t^2 - 14t + 8 ?$$

Derive also the particle's velocity function. What is the value of its acceleration?

18. A body's motion is described by $x = 2t^3 - t^2 + 3t - 7$. Find its speed, velocity and acceleration when $t = 1$.

19. An object's position along the x-axis is described by the equation $x = 5t^3 - 20t^2 + 25t - 6$. At what time will the object be at rest? What is its acceleration and displacement at this time?

20. What is the total distance traversed by a body in the first 6 seconds of motion if its displacement function is given by $x = t^3 - 8t^2 + 20t - 50$ metres.

21. The equation of motion for a particle is given by $x = t^3 + 12t^2 + 36t - 10$. At what times is the particle instantaneously at rest? What is its acceleration at these times?

22. The position of a small mass for an experiment is described by $s = 14t - t^3 + 20$, $0 < t < 4$. Find the displacement at the moment the velocity is zero.

23. An object moving vertically has its motion described by the equation $y = 30t - 18t^2 + 1$. Find the object's initial speed and starting height. Also find the distance covered by the object between the two moments it is at its starting height.

24. A ball thrown straight up in the air reaches a height of h metres in t seconds. The height is determined by the equation $h = 25t - 5t^2$. What is the ball's speed after 2 seconds? What is the maximum height attained? Determine the acceleration of the ball.

25. A body moving along the x-axis according to the x—t relationship $x = \frac{1}{3}t^3 - \frac{9}{2}t^2 + 8t + 2$, will be at rest at certain times. What is the body's distance and direction from the origin at these times? What is the body's acceleration at these times?

26. A particle moves according to the equation of motion $x = -\frac{1}{4}t^4 + 6t^3 - \frac{95}{2}t^2 + 150t + 12$. At what times will the velocity be zero?

27. The side of a square floor tile is 203 mm, however its area is taken to be only 400 cm^2. Find the absolute and percentage error in the area using calculus techniques.

28. Find the approximate change in the value of the function $7x^2 - 10x + 5$ as x changes from 2 to 2.001.

29. The distance in metres of a moving body in time t seconds is given by the formula $s = 20t - 5t^2$. Find the approximate distance through which the body moves in one-tenth of a second after 5 seconds.

30. A coat of paint of thickness x mm is applied to a cylindrical building column of radius a mm and height h m. Find the exact amount of paint used by determining the volumes both before and after painting. Using differentiation now, what is the approximate amount of paint used?

31. It is known that the formula for the period of a pendulum is approximately given by

$$T = 2\pi \sqrt{\frac{L}{g}}$$

where L is the pendulum's length and g the Earth's gravitational constant. If the length of the pendulum is increased by some small amount $a\%$, determine the percentage increase in the period of the swing of the pendulum.

32. In chemistry, Boyle's law states that the product of the pressure P and volume V of a gas is a constant. i.e. $PV = k$. Express the relative change in the pressure of a gas in terms of the relative change in the volume, using ΔP and ΔV to represent small changes in the pressure and volume respectively.

33. In astrophysics the luminosity of a star can be determined using the Stefan-Boltzmann law,

$$L = (4\pi R^2)(\sigma T^4)$$

where R is the radius of the star, σ the Stefan-Boltzmann constant and T the temperature of the stellar surface. Find the approximate percentage increase in the luminosity of the Sun should its surface temperature increase by 1%. The radius of the Sun is 7×10^8 m with surface temperature 5800 K and luminosity of 3.85×10^{26} W. Stefan-Boltzmann constant equals 5.67×10^{-8}.

34.

a) Find the angles that the tangents to the function $f(x) = 5x - x^2$ makes with the x-axis.

b) Find the co-ordinates of the points where the curve $y = 2(x - 2)(x + 1)(x + 3)$ intersects the x-axis. At what angles do the tangents of the curve at these points make with the x-axis.

c) Find the equation of the tangent to the parabola $y = 6x - x^2$ at the point $(-1, 2)$.

35. A projectile encounters a highly viscous medium and penetrates to a depth of x cm given by $x = 9t - 0.5t^3$. Determine the distance the projectile will move into the medium before coming to a stop.

36. Find the equation of the lines running parallel to the line $y = 3x - 1$ that lie tangent to the curve $5y = x^3 - 2x$.

37. A cyclist is riding up hilly terrain that has a profile approximated by the function

$$h = \frac{1}{20800}x^3 - \frac{21}{2080}x^2 + \frac{15}{26}x$$

over the first 150 metres. Both height h and horizontal distance x are measured in metres. Taking the origin as the cyclist's starting point, find the average change in the height over the first 150 metres. Also, find the distance from his starting point where he can expect to find the same gradient as that he begins his ride with.

Answers

1) $y = -6x + 8$
2) $3x + 17$
3) $x + 3y = -1$
4) $10x + y = -7; 10y - x = 31$
5) $y = 4x - 13; x + 4y = 16$
6) $\frac{3}{5}$
7) $20\frac{5}{12}$
8) $\left(\frac{-1+\sqrt{7}}{2}, \frac{12-3\sqrt{7}}{2}\right); \left(\frac{-1-\sqrt{7}}{2}, \frac{12+3\sqrt{7}}{2}\right); 2y = 6(3\sqrt{7} - 1)x + 9(\sqrt{7} - 6); 3(1 + \sqrt{7})x + 2y = -3(4 + \sqrt{7}); (3 - \sqrt{7})x + y = 1 + \frac{1}{2}\sqrt{7}; (3 + \sqrt{7})x + y = 1 - \frac{1}{2}\sqrt{7}$
9) $a = -1, b = -7, c = 3$
10) $2x + 2y = 5$
11) $b = -\frac{1}{16c}$
12) $\left(-3\frac{2}{3}, -128\frac{1}{3}\right)$
13) $6y - 9x = 36 + \frac{1}{\sqrt{6}}; 6y - 9x = 36 - \frac{1}{\sqrt{6}}$
14) $x + y = 2; (2,0)$

15) $a = -\frac{28}{81}, b = 3\frac{56}{81}, c = 1\frac{28}{81}$

16) $0,53\frac{1}{4}, -437\frac{1}{3}$

17) $8; 6t - 14; 6$

18) $1; -1; 2$

19) $\frac{5}{3}, 1; 10, -10; 9\frac{1}{27}, 4$

20) $50\frac{10}{27}$ m

21) $2, 6; 12$

22) $2\sqrt[3]{\frac{14}{3}} + 20$

23) 25

24) 5 m/s; 31.25 m; -10 m/s^2

25) $5\frac{5}{6}$ units to right of origin, $51\frac{1}{3}$ units to right of origin; -7, 7

26) $3, 5, 10$

27) 12.18 cm^2, 2.96%

28) 0.018

29) 3 m

30) $1000\pi hx(2a + x)$ mm^3; $2000\pi ahx$ mm^3

31) $0.5a\%$

32) $-\frac{\Delta V}{V}$

33) 4%

34) a) $78.7°, 101.3°$
   b) $(-3, 0); (-1, 0); (2, 0); 87.1°, 94.8°, 88.1°$
   c) $y = 8x + 10$

35) $6\sqrt{6}$ cm

36) $45\sqrt{3}x - 15\sqrt{3}y = 34\sqrt{17}; 15\sqrt{3}y - 45\sqrt{3}x = 22\sqrt{17}$

37) 0.144, 140 m

Chapter 3

The Rules of Differentiation

3.1 Introduction

The inventors of the differential calculus, Newton and Leibniz, soon discovered rules for the differentiation of the product, $f(x).g(x)$, and the quotient, $f(x)/g(x)$, of two functions $f(x)$ and $g(x)$ whose derivatives are known. They also found rules for finding the derivative of functions raised to a power, i.e. $[f(x)]^n$ and for composite functions $f(g(x))$.

In Chapter 1 we saw how these mathematicians first discovered the rule for differentiating polynomial expressions. We shall apply the same limiting process to obtain simplified general rules for differentiating more complex expressions.

3.2 The Derivative of a Product $f(x).g(x)$

We wish to find a rule for taking the derivative for a product of functions. Using the Fermat-Newton definition of the derivative we have

$$\frac{d}{dx}[f(x) \cdot g(x)] = \lim_{h \to 0} \left[\frac{f(x+h)g(x+h) - f(x)g(x)}{h} \right]$$

and

$$\lim_{h \to 0} \left[\frac{f(x+h) - f(x)}{h} \right] = \frac{df}{dx}$$

$$\lim_{h \to 0} \left[\frac{g(x+h) - g(x)}{h} \right] = \frac{dg}{dx}$$

This means, for h very small, we can write

$$f(x+h) - f(x) = h \cdot \frac{df}{dx}$$

$$g(x+h) - g(x) = h \cdot \frac{dg}{dx}$$

We can therefore write

$$f(x+h) \cdot g(x+h) = \left[f(x) + h \cdot \frac{df}{dx} \right] \left[g(x) + h \cdot \frac{dg}{dx} \right]$$

i.e.

$$f(x+h) \cdot g(x+h) = f(x) \cdot g(x) + h \left[f(x) \cdot \frac{dg}{dx} + g(x) \cdot \frac{df}{dx} \right]$$

$$+ h^2 \cdot \frac{df}{dx} \frac{dg}{dx}$$

Therefore,

$$f(x+h) \cdot g(x+h) - f(x) \cdot g(x)$$

$$= h \left[f(x) \cdot \frac{dg}{dx} + g(x) \cdot \frac{df}{dx} \right] + h^2 \cdot \frac{df}{dx} \cdot \frac{dg}{dx}$$

So, we find that

$$\frac{d}{dx}[f(x) \cdot g(x)] = \lim_{h \to 0} \left[\frac{h \left(f(x) \cdot \frac{dg}{dx} + g(x) \cdot \frac{df}{dx} + h^2 \cdot \frac{df}{dx} \cdot \frac{dg}{dx} \right)}{h} \right]$$

$$= \lim_{h \to 0} \left[f(x) \cdot \frac{dg}{dx} + g(x) \cdot \frac{df}{dx} \right] + \lim_{h \to 0} \left[h \cdot \frac{df}{dx} \cdot \frac{dg}{dx} \right]$$

i.e.

$$\frac{d}{dx}[f(x) \cdot g(x)] = f(x) \cdot \frac{dg}{dx} + g(x) \cdot \frac{df}{dx}$$

With an even simpler notation we could write

$$\frac{d}{dx}(uv) = uv' + vu'$$

The proof that we have just given applies to all functions $f(x), g(x)$ for which the derivative curves $df/dx, dg/dx$ exist at the points at which we want to determine $\frac{d}{dx}[f(x) \cdot g(x)]$. We call such functions *differentiable*. It will suffice our purposes here to note simply that a differentiable function is one whose curve has only one tangent at each point. For example, an absolute value function has an infinite number of tangents that can be drawn at its 'sharp-pointed' vertex, and would therefore not be considered differentiable.

Our proof of the law for the differentiation of the product of two functions can be given a geometrical form as shown in Fig. 3.1. Thinking of $f(x) . g(x)$ as an *area,* we see that the change in area when we go $x \rightarrow x + h$ is given by

$$h\left[f(x)\frac{dg}{dx} + g(x)\frac{df}{dx}\right] + h^2\frac{df}{dx} \cdot \frac{dg}{dx}$$

i.e.

$$f(x + h)g(x + h) - f(x)g(x)$$

$$= h\left[f(x)\frac{dg}{dx} + g(x)\frac{df}{dx}\right] + h^2\frac{df}{dx} \cdot \frac{dg}{dx}$$

Therefore,

$$\lim_{h \to 0} \frac{f(x + h)g(x + h) - f(x)g(x)}{h}$$

$$= f(x)\frac{dg}{dx} + g(x)\frac{df}{dx} + \lim_{h \to 0} h \cdot \frac{df}{dx} \cdot \frac{dg}{dx}$$

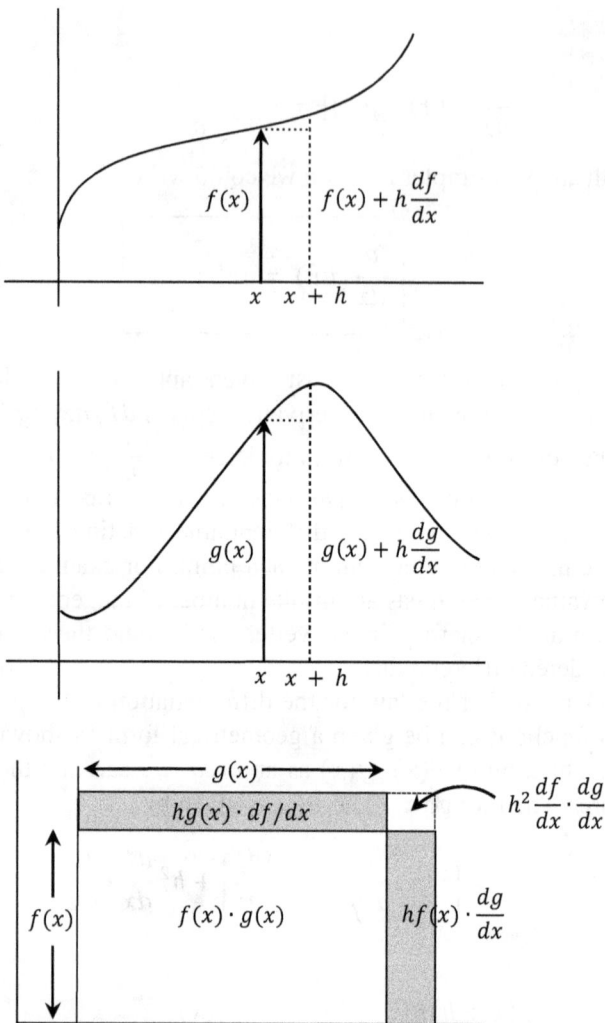

Fig. 3.1. A Geometrical proof of the product rule.

That is, as found before,

$$\frac{d}{dx}[f(x) \cdot g(x)] = f(x)\frac{dg}{dx} + g(x)\frac{df}{dx}$$

Example 3.1. *If* f *is the function defined by* $f(x) = (3x + 5)(4x^2 - x + 7)$, *find* $f'(x)$.

Let $u = 3x + 5$ and $v = 4x^2 - x + 7$. So,

$$f'(x) = u \cdot \frac{dv}{dx} + v \cdot \frac{du}{dx}$$

$$= (3x + 5) \cdot \frac{d}{dx}(4x^2 - x + 7) + (4x^2 - x + 7)$$

$$\cdot \frac{d}{dx}(3x + 5)$$

$$= (3x + 5)(8x - 1) + (4x^2 - x + 7) \cdot 3$$

$$= 24x^2 - 3x + 40x - 5 + 12x^2 - 3x + 21$$

$$= 36x^2 + 34x + 16$$

$\square$

Note that these types of problems can also be solved by expanding the brackets followed by differentiation of the resulting polynomial, as shown in example 2.7.

Example 3.2. *If* $f(x) = (3x^2 - 5)(7x^3 + 6x + 1)$, *find* $f'(x)$.

We have,

$$f'(x) = (3x^2 - 5) \cdot \frac{d}{dx}(7x^3 + 6x + 1) + (7x^3 + 6x + 1)$$

$$\cdot \frac{d}{dx}(3x^2 - 5)$$

$$= (3x^2 - 5)(21x^2 + 6) + (7x^3 + 6x + 1)(6x)$$

$$= 63x^4 + 18x^2 - 105x^2 - 30 + 42x^4 + 36x^2 + 6x$$

$$= 105x^4 - 51x^2 + 6x - 30$$

Alternatively, by expanding the brackets in the given function first,

$$f(x) = 21x^5 + 18x^3 + 3x^2 - 35x^3 - 30x - 5$$

$$= 21x^5 - 17x^3 + 3x^2 - 30x - 5$$

Differentiation gives,

$$f'(x) = 105x^4 - 51x^2 + 6x - 30$$

as before.

□

3.3 The Derivative of Positive Powers of $f(x)$

As a simple extension to the use of the product rule, let us see what we obtain when we set $g(x) = f(x)$. We then find

$$\frac{d}{dx}[f(x) \cdot g(x)] = \frac{d}{dx}[f^2(x)]$$

$$= f(x) \cdot \frac{df}{dx} + f(x) \cdot \frac{df}{dx}$$

$$= 2f(x) \cdot \frac{df}{dx}$$

Similarly, setting

$$g(x) = f^2(x)$$

we find

$$\frac{d}{dx}[f(x) \cdot g(x)] = \frac{d}{dx}[f^3(x)]$$

$$= f(x) \cdot \frac{d}{dx}[f^2(x)] + f^2(x) \cdot \frac{df}{dx}$$

Utilising the previous result we have,

$$\frac{d}{dx}[f^3(x)] = f(x) \cdot 2f(x) \cdot \frac{df}{dx} + f^2(x) \cdot \frac{df}{dx} = 3f^2(x) \cdot \frac{df}{dx}$$

In the same way we find

$$\frac{d}{dx}[f^4(x)] = \frac{d}{dx}[f(x) \cdot f^3(x)]$$

$$= f(x) \cdot \frac{d}{dx}[f^3(x)] + f^3(x) \cdot \frac{df}{dx}$$

$$= 3f^3(x) \cdot \frac{df}{dx} + f^3(x) \cdot \frac{df}{dx}$$

$$= 4f^3(x) \cdot \frac{df}{dx}$$

and generally that

$$\frac{d}{dx}(f^n(x)) = nf^{n-1}(x) \cdot \frac{df}{dx}$$

We therefore see that we can differentiate powers of a function $f(x)$ just as if they were simple powers of x, as long as we remember to multiply by a term df / dx.

This formula is one form of what is known as the *chain rule*. We shall revisit the chain rule later in this chapter.

Example 3.3. *Find the derivative of the polynomial function defined by*

$$f(x) = (5x + 3)^3$$

Solution:

$$f'(x) = 3(5x + 3)^2 \cdot \frac{d}{dx}(5x + 3)$$

$$= 3(5x + 3)^2(5)$$

$$= 15(5x + 3)^2$$

□

Example 3.4. *Take the derivative of*

$$y = (7x^2 + 3)^4 - 3x^2 + 1.$$

Solution:

$$\frac{dy}{dx} = 4(7x^2 + 3)^3(14x) - 6x$$

$$= 56x(7x^2 + 3)^3 - 6x$$

$$= 2x[28(7x^2 + 3)^3 - 3]$$

□

We now have simple rules for differentiating products as well as powered functions. We would also like to have a simple formula for finding the derivative of a quotient. This can be accomplished by extending the previous result to *negative* powers of $f(x)$.

3.4 The Derivative of Negative Powers of $f(x)$

We would now like to find the derivative of the reciprocal of $f(x)$, the function $1/f(x)$. We use a trick to achieve this. Consider the function

$$F(x) = \frac{f^2(x)}{f(x)} = f(x)$$

Differentiating this function using our rule for the differentiation of a product we find

$$\frac{dF}{dx} = \frac{d}{dx}\left(f^2(x) \cdot \frac{1}{f(x)}\right)$$

$$= f^2(x) \cdot \frac{d}{dx}\left(\frac{1}{f(x)}\right) + \frac{1}{f(x)} \cdot \frac{d}{dx}(f^2(x))$$

i.e.

$$\frac{dF}{dx} = \frac{df}{dx} = f^2(x) \cdot \frac{d}{dx}\left(\frac{1}{f(x)}\right) + \frac{1}{f(x)} \cdot 2 \cdot f(x) \cdot \frac{df}{dx}$$

i.e.

$$\frac{df}{dx} - 2\frac{df}{dx} = f^2(x) \cdot \frac{d}{dx}\left(\frac{1}{f(x)}\right) - \frac{df}{dx}$$

$$= f^2(x) \cdot \frac{d}{dx}\left(\frac{1}{f(x)}\right)$$

So,

$$\frac{d}{dx}\left(\frac{1}{f(x)}\right) = -\frac{1}{f^2(x)} \cdot \frac{df}{dx}$$

In the same way by considering

$$\frac{d}{dx}\left(f^3(x) \cdot \frac{1}{f^2(x)}\right) = f^3(x) \cdot \frac{d}{dx}\left(\frac{1}{f^2(x)}\right)$$

$$+ \frac{1}{f^2(x)} \cdot \frac{d}{dx}(f^3(x))$$

i.e.

$$\frac{df}{dx} = f^3(x) \cdot \frac{d}{dx}\left(\frac{1}{f^2(x)}\right) + \frac{1}{f^2(x)} \cdot 3f^2(x) \cdot \frac{df}{dx}$$

$$= f^3(x) \cdot \frac{d}{dx}\left(\frac{1}{f^2(x)}\right) + 3 \cdot \frac{df}{dx}$$

i.e.

$$-2\frac{df}{dx} = f^3(x) \cdot \frac{d}{dx}\left(\frac{1}{f^2(x)}\right)$$

we find

$$\frac{d}{dx}\left(\frac{1}{f^2(x)}\right) = -\frac{2}{f^3(x)} \cdot \frac{df}{dx}$$

In general by setting

$$\frac{d}{dx}\left(f^{n+1}(x) \cdot \frac{1}{f^n(x)}\right) = f^{n+1}(x) \cdot \frac{d}{dx}\left(\frac{1}{f^n(x)}\right)$$

$$+ \frac{1}{f^n(x)} \cdot \frac{d}{dx}(f^{n+1}(x))$$

we find

$$\frac{df}{dx} = f^{n+1}(x) \cdot \frac{d}{dx}\left(\frac{1}{f^n(x)}\right) + \frac{n+1}{f^n(x)} \cdot f^n(x) \cdot \frac{df}{dx}$$

so that

$$\frac{d}{dx}\left(\frac{1}{f^n(x)}\right) = -\frac{n}{f^{n+1}(x)} \cdot \frac{df}{dx} \qquad (3.1)$$

Example 3.5. *Find the derivative of* $1/(x + 1)^5$.

Solution:

Let

$$f(x) = \frac{1}{(x + 1)^5}$$

Now,

$$f'(x) = -\frac{5}{(x + 1)^{5+1}} \cdot \frac{d}{dx}(x + 1) = -\frac{5}{(x + 1)^6}$$

□

Example 3.6. *Find* df / dx *for* $f(x) = 1/(x^2 - 3)^3$.

Solution:

$$f'(x) = -\frac{3}{(x^2 - 3)^4} \cdot \frac{d}{dx}(x^2 - 3)$$

$$= -\frac{3}{(x^2 - 3)^4} \cdot 2x$$

$$= -\frac{6x}{(x^2 - 3)^4}$$

□

As an alternative to the method used in these last two examples, we could also have rewritten the function with a negative power and applied the chain rule as we did in the previous section. For example, the previous example could be performed in the following way:

$$f(x) = (x^2 - 3)^{-3}$$

So,

$$f'(x) = -3(x^2 - 3)^{-4} \cdot 2x$$

$$= -6x(x^2 - 3)^{-4}$$

$$= \frac{-6x}{(x^2 - 3)^4}$$

3.5 Derivative of a Quotient $f(x) / g(x)$

Utilising equation 3.1 we can now form the derivative of a quotient $f(x) / g(x)$. We have

$$\frac{d}{dx}\left(\frac{f(x)}{g(x)}\right) = f(x) \cdot \frac{d}{dx}\left(\frac{1}{g(x)}\right) + \frac{1}{g(x)} \cdot \frac{d}{dx}(f(x))$$

Now

$$\frac{d}{dx}\left(\frac{1}{g(x)}\right) = -\frac{1}{g^2(x)} \cdot \frac{dg}{dx}$$

so that

$$\frac{d}{dx}\left(\frac{f(x)}{g(x)}\right) = -\frac{f(x)}{g^2(x)} \cdot \frac{dg}{dx} + \frac{1}{g(x)} \cdot \frac{df}{dx}$$

or,

$$\frac{d}{dx}\left(\frac{f(x)}{g(x)}\right) = \frac{g(x) \cdot \frac{df}{dx} - f(x) \cdot \frac{dg}{dx}}{g^2(x)}$$

Using Lagrange's notation with u and v to represent $f(x)$ and $g(x)$, gives a simpler form of the *quotient rule*:

$$\boxed{\frac{d}{dx}\left(\frac{u}{v}\right) = \frac{vu' - uv'}{v^2}}$$

Example 3.7. *Find $f'(x)$ given $f(x) = \frac{2x-5}{x^2+6}$.*

Let $u = 2x - 5$ and $v = x^2 + 6$. So, $u' = 2$ and $v' = 2x$. Therefore, the quotient rule gives:

$$f' = \frac{(x^2 + 6) \cdot 2 - (2x - 5) \cdot 2x}{(x^2 + 6)^2}$$

$$= \frac{2x^2 + 12 - 4x^2 + 10x}{(x^2 + 6)^2}$$

$$= \frac{-2x^2 + 10x + 12}{(x^2 + 6)^2}$$

$$= \frac{-2(x^2 - 5x - 6)}{(x^2 + 6)^2}$$

$$= -\frac{2(x - 6)(x + 1)}{(x^2 + 6)^2}$$

□

Example 3.8. *Given*

$$f(x) = \frac{3x^4 - 7x + 8}{6x^3 + 1}$$

find $f'(x)$.

Applying the quotient rule directly we have,

$$f'(x) = \frac{(6x^3 + 1)(12x^3 - 7) - (3x^4 - 7x + 8)(18x^2)}{(6x^3 + 1)^2}$$

$$= \frac{72x^6 - 42x^3 + 12x^3 - 7 - 54x^6 + 126x^3 - 144x^2}{(6x^3 + 1)^2}$$

$$= \frac{18x^6 + 96x^3 - 144x^2 - 7}{(6x^3 + 1)^2}$$

□

Example 3.9. *If*

$$y = \frac{(x + 1)^3}{(x^2 - 3)^2}$$

find $\frac{dy}{dx}$.

We have

$$\frac{dy}{dx} = \frac{(x^2 - 3)^2 \cdot 3(x + 1)^2 - (x + 1)^3 \cdot 2(x^2 - 3) \cdot 2x}{[(x^2 - 3)^2]^2}$$

$$= \frac{(x^2 - 3)(x + 1)^2[3(x^2 - 3) - 4x(x + 1)]}{(x^2 - 3)^4}$$

$$= \frac{(x + 1)^2(-x^2 - 4x - 9)}{(x^2 - 3)^3}$$

$$= -\frac{(x + 1)^2(x^2 + 4x + 9)}{(x^2 - 3)^3}$$

$$= \frac{(x + 1)^2(x^2 + 4x + 9)}{(3 - x^2)^3}$$

$\square$

We will now consider the problem of finding the derivative of the composite function $f(g(x))$.

3.6 The Derivative of the Composite Function $f(g(x))$

From curve II in Fig. 3.2 below we see that the gradient of the tangent line can be written,

$$\frac{df}{dg} = \frac{\Delta f + \varepsilon_2}{\Delta g}$$

from which we can write,

$$\Delta f = \frac{df}{dg}\Delta g - \varepsilon_2$$

Dividing by Δx gives,

$$\frac{\Delta f}{\Delta x} = \frac{df}{dg} \cdot \frac{\Delta g}{\Delta x} - \frac{\varepsilon_2}{\Delta x} \tag{3.1}$$

By similar working, from curve I in Fig. 3.2 we can write

$$\frac{\Delta g}{\Delta x} = \frac{dg}{dx} - \frac{\varepsilon_1}{\Delta x} \tag{3.2}$$

Substituting equation (3.2) into equation (3.1) gives

$$\frac{\Delta f}{\Delta x} = \frac{df}{dg}\left(\frac{dg}{dx} - \frac{\varepsilon_1}{\Delta x}\right) - \frac{\varepsilon_2}{\Delta x} \tag{3.3}$$

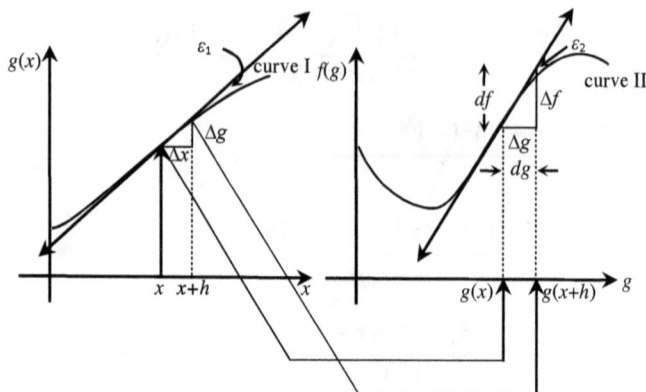

Fig. 3.2. Determining derivative of a composite function.

We now want to consider the behaviour of (3.3) as $h \rightarrow 0$, i.e. as $\Delta x \rightarrow 0$. Since the ε's are being 'squeezed' between the tangent and the curve, they diminish at a greater rate than does Δx as $\Delta x \rightarrow 0$, resulting in $\frac{\varepsilon_1}{\Delta x}, \frac{\varepsilon_2}{\Delta x} \rightarrow 0$. Thus, (3.3) becomes

$$\frac{df}{dx} = \frac{df}{dg} \cdot \frac{dg}{dx}$$

This expression is *also* known as the *chain rule*, though is of a different form to what we saw earlier in this chapter. In the Lagrange notation it can be written

$$f\big(g(x)\big)' = g'(x)f'(g(x))$$

Example 3.10. *Find the derivative of the composite function* $f(g(x))$ *where* $g(x) = x^2$ *and* $f(x) = x(1 - x)$.

Replacing x in the f function with the g function we have

$$f\big(g(x)\big) = g(1 - g) = g - g^2$$

Deriving gives

$$\frac{d}{dx}[f(g(x))] = \frac{d}{dx}[g(x)] \cdot \frac{d}{dg}[f(g)]$$

$$= \frac{d}{dx}(x^2) \cdot \frac{d}{dg}(g - g^2)$$

$$= 2x(1 - 2g)$$

$$= 2x(1 - 2x^2)$$

$$= 2x - 4x^3$$

□

Example 3.11. *Given the two functions $f(x) = x + 10$ and $g(x) = (x - 1)^2/(x + 2)^3$, find $\frac{d}{dx}[f(g(x))]$.*

Solution:

$$f(g(x)) = g + 10 = \frac{(x - 1)^2}{(x + 2)^3} + 10$$

$$\frac{d}{dx}[f(g(x))] = \frac{d}{dx}\left[\frac{(x - 1)^2}{(x + 2)^3}\right] \cdot \frac{d}{dg}[g + 10]$$

$$= \frac{(x + 2)^3 \cdot 2(x - 1) - (x - 1)^2 \cdot 3(x + 2)^2}{[(x + 2)^3]^2} \cdot 1$$

$$= \frac{(x - 1)(x + 2)^2[2(x + 2) - 3(x - 1)]}{(x + 2)^6}$$

$$= \frac{(x - 1)(7 - x)}{(x + 2)^4}$$

□

Note that within this solution we have used our earlier *chain rule* for differentiating expressions in the form $(ax + b)^n$, where $f(x) = x^n$ and $g(x) = ax + b$. As expressions become increasingly complex then it is common to be applying a number of differentiation rules, one inside the other, as it were.

3.7 The Derivative of Parametric Equations

Sometimes it is convenient to relate two quantities via an inter-mediary—the *parametric*. For example, in circular motion the position of a point in the Cartesian plane can be plotted by set-ting $x = \sin(t)$ and $y = \cos(t)$, with t representing time. Finding the gradient function for the circle may be found using the *chain rule*.

In determining the gradient function for parametric equations, it is often convenient to take the derivative with respect to the independent variable, but then utilise the reciprocal. For exam-ple, we may be requiring the use of dt/dy, but the given equa-tion is explicit with t. That is, dt/dy may be difficult to find but its reciprocal dy/dt relatively easy.

The chain rule itself can be used to prove that $\dfrac{dt}{dy} = \dfrac{1}{dy/dt}$. From the chain rule we can write

$$\frac{dy}{dx} \cdot \frac{dx}{dy} = \frac{dy}{dy} = 1$$

Therefore,

$$\boxed{\frac{dx}{dy} = \frac{1}{dy/_{dx}}, \quad \frac{dy}{dx} \neq 0}$$

assuming x and y are differentiable functions.

In finding the gradient function of parametric equations, take the derivative of each equation with respect to t, and then use the chain rule to find the derivative with respect to the independent variable x. The following example will illustrate this.

Example 3.12. *Find the gradient function of the curve repre-sented by the parametric equations* $x = 5t^2$, $y = 3t$.

Differentiating each parametric equation with respect to t gives

$$\frac{dx}{dt} = 10t, \quad \frac{dy}{dt} = 3$$

Therefore, by the chain rule,

$$\frac{dy}{dx} = \frac{dy}{dt} \cdot \frac{dt}{dx} = 3 \cdot \frac{1}{10t} = \frac{3}{10t}$$

Since from $x = 5t^2$ we have

$$t^2 = \frac{x}{5}$$

$$t = \pm\sqrt{\frac{x}{5}}$$

then,

$$\frac{dy}{dx} = \frac{3}{10 \cdot \pm\sqrt{\frac{x}{5}}} = \pm\frac{3}{2\sqrt{5x}}$$

□

Sometimes it is insightful to plot the curve described by the parametric equations. One method in doing so is to simply remove the parametric variable by performing simultaneous equations and graphing the resulting function or relation. For example, using the equations from example 3.12 above we have,

$$t = \frac{1}{3}y$$

Substituting into the x equation gives

$$x = 5\left(\frac{1}{3}y\right)^2 = \frac{5}{9}y^2$$

which represents a parabola aligned along the x-axis with turning point at the origin. Alternatively a range of t-values could be substituted into the parametric equations obtaining the x, y-coordinates of points along the curve. For example, see table 3.1 below.

Plotting the (x, y) points and drawing a smooth curve connecting them results in the parabola shown in Fig. 3.3. Parametric curves have a direction of motion, so we include arrows pointing in the direction of increasing t.

t	x	y
-2	20	-6
-1.5	11.25	-4.5
-1	5	-3
-0.5	1.25	-1.5
0	0	0
0.5	1.25	1.5
1	5	3
1.5	11.25	4.5
2	20	6

Table 3. 1. Determining coordinates from parametric input.

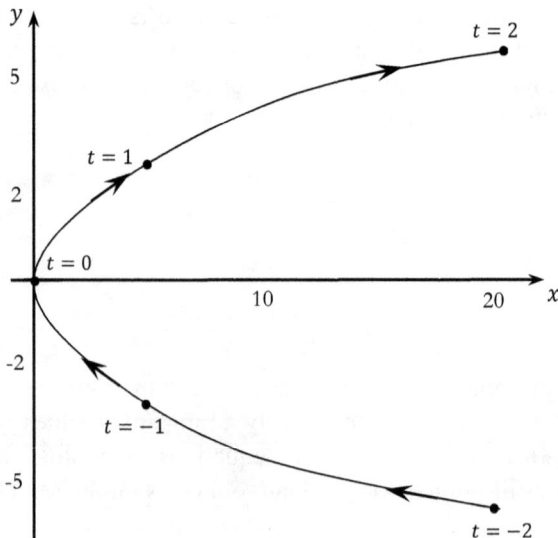

Fig. 3.3. Showing the graph of a parametric curve.

Exercises 3

1. Differentiate the following expressions with respect to x.

   (i) $(2x + 3)(x + 4)$ (ii) $\left(5x - \frac{2}{x}\right)(6x + 7)$

   (iii) $(3x^2 + 0.5x - 11)(4 - 15x)$

   (iv) $(x + 1)^2(3x - 2)^3$ (v) $\left(\frac{1}{2}x - x^{-1}\right)^4 (2x - 1)^6$

   (vi) $\frac{3x^2 - 4}{x^3 + 2}$

2. Determine the first and second derivative of

$$y = (6x^3 + 2x^2 - 5x + 9)^4$$

3. Determine $\frac{d^2y}{dx^2}$ and find its value(s) when $y = -1$ for $y = \frac{x^2 + 1}{x - 3}$.

4. Differentiate with respect to x the following expressions:

   (i) $\frac{x^2 - 7x + 1}{(3x^2 - 5)^2}$ (ii) $\sqrt{x^3 + \sqrt{2x^2 + 1}}$ (iii) $\sqrt{x\sqrt{x\sqrt{x}}}$

   (iv) $\sqrt[3]{3x - 7}$

5. Find $f'(x)$ where $f(x)$ equals the following expressions:

   (i) $(3x - 5)(x + 7)(6x - 11)$ (ii) $x^5\sqrt{2x^3 + 3}$

   (iii) $(5x - 4)\sqrt{15x - 1}$ (iv) $\sqrt{3x + 1}\sqrt{2 - 5x}$

6. Find the derivative of the following functions:

   (i) $f(x) = \frac{(x + 1)^3}{x^2 - 3x + 6}$

   (ii) $g(x) = \frac{3x}{\sqrt[3]{x - 2}}$

   (iii) $h(x) = \frac{x^2\sqrt{x - 4}}{x + 10}$

   (iv) $A(x) = \sqrt{\frac{x/3 - 2}{x^2 - 2}}$

(v) $B(x) = \dfrac{(3x-1)^2(x+5)^5}{\sqrt{x-2}}$

(vi) $p(x) = \dfrac{x^3 - 3x^2 + 2}{2x^3 + x^2 + 6x}$

7. Differentiate:

(i) $(2x^3 + 4x^2 - 3x + 1)^7 (9x^2 + 5x - 6)^3$

(ii) $\sqrt[4]{\sqrt{x^3 + 1}}\sqrt{2x - 8}$

(iii) $\dfrac{(x^2 - 9)(x^2 + 6x + 8)}{(x+3)(x+2)(x+4)}$

(iv) $\dfrac{\left(\sqrt{3}x^2 + \frac{1}{2}\right)^2 (2x^3 + 3)^3}{\left(\frac{1}{4}x^2 - 7\right)^{\frac{1}{2}}}$

8. Find the equation of the tangent to the curve $x = t^2, y = 4t - t^3$ at the point where $t = 2$.

9. Find the equation of the normal to the curve $x = 2t^2, y = t^3 - 3t$ at the point where $t = 3$.

10. A body's position after t seconds is given by the parametric equations $x = 12t$, $y = 30t - 8t^2$. Find the gradient of the tangent the body is moving along at time t and the time when the motion is horizontal.

11. Graph the curve defined by the parametric equations

$$x = 3t^5 - 6t^3 \text{ and } y = 2t^2$$

indicating direction of motion. Also, find the equation(s) of the tangent(s) to the curve at the point (0, 4).

12. At which points does the parametric curve, described by equations $x = t^3 - 2t$ and $y = 2t^2 - 8$, have horizontal and vertical tangents?

13. Without any working, state the derivative of $g(5x^2)$ given

$$\frac{d}{dx}[g(x)] = \frac{1}{\sqrt[3]{1 - x^2}}$$

14. The parametric equations of the foliums of Descartes are given by

$$x(t) = \frac{3at}{t^3 + 1}, \quad y(t) = \frac{3at^2}{t^3 + 1}$$

where a is a constant. Draw a sketch of the curve and find the equation of the tangent to the curve for $a = 3$ at the point (2, 4).

Answers

1) i) $4x + 11$ ii) $60x + 35 + 14/x^2$ iii) $9x - 135x^2 + 167$
iv) $5(x + 1)(3x + 1)(3x - 2)^2$

v) $(2x - 1)^5 \left(\frac{1}{2}x - \frac{1}{x}\right)^3 \left(10x - \frac{4}{x} - \frac{4}{x^2} - 2\right)$ vi) $\frac{3x(4x - x^3 + 4)}{(x^3 + 2)^2}$

2) $4(6x^3 + 2x^2 - 5x + 9)^2[3(18x^2 + 4x - 5)$
$\qquad\qquad + 4(9x - 1)(6x^3 + 2x^2 - 5x + 9)]$

3) $\frac{20}{(x-3)^2}; \frac{4}{5}$ 4) i) $\frac{6x^3 - 63x^2 + 22x - 35}{(5 - 3x^2)^3}$ ii) $\frac{x\left(3x\sqrt{2x^2+1}+2\right)}{2\sqrt{(2x^2+1)(x^3+\sqrt{2x^2+1}}}$

iii) $\frac{7}{8\sqrt[8]{x}}$ iv) $\frac{1}{\sqrt[3]{(3x-7)^2}}$ 5) i) $54x^2 + 126x - 386$

ii) $2x^4(7x^3 + 6)$ iii) $\frac{5(45x-14)}{2\sqrt{15x-1}}$ iv) $\frac{1-30x}{2\sqrt{(3x+1)(5x-2)}}$

6) i)$(x + 1)^2(x^2 - 8x + 21)$ ii) $\frac{2(x-3)}{\sqrt[3]{(x-2)^4}}$ iii) $\frac{x(3x^2+42x-160)}{2\sqrt{x-4}(x+10)^2}$

iv) $-\frac{1}{6}\left(\frac{3x^2-2}{x-6}\right)^{\frac{2}{3}}\left(\frac{x^2-12x+2}{(x^2-2)^2}\right)$ v) $\frac{(3x-1)(x+5)^4(39x^2-48x-95)}{2(x-2)^{\frac{3}{2}}}$

vi) $\frac{3x^2(2x^2+x+6)(x-2)-2(x-1)(x^2-2x-2)(3x^2+x+3)}{x^2(2x^2+x+6)^2}$

7) i) $(9x^2 + 5x - 6)^2(2x^3 + 4x^2 - 3x + 1)^6(486x^4 + 960x^3$
$\qquad\qquad - 263x^2 - 537x + 141)$

ii) $\frac{\sqrt{2}(7x^3-12x^2+4)}{8\cdot\sqrt[4]{\sqrt{(x^3+1)^7(x-4)^2}}}$ iii) 1

iv) $\frac{x(2x^2+3)(2\sqrt{3}x^2+1)[(28-x^2)(4\sqrt{3}x^5+2(9\sqrt{3}-1)x^3+2\sqrt{3}x^2+9x+3(4\sqrt{3}-1)]}{2(x^2-28)^{\frac{3}{2}}}$

8) $x + y = 8$ 9) $x + 2y = 54$ 10)$\frac{15-8t}{6}; \frac{15}{8}s$ 11) $6y \pm \sqrt{2}x = 24$

12) $\left(\pm\frac{4}{3}\sqrt{\frac{2}{3}}, -5\frac{17}{27}\right), (0, -8)$ 13) $\frac{10x}{\sqrt[3]{1-25x^4}}$ 14) $5y - 4x = 12$

Chapter 4

Differentiation of Trigonometric Functions

4.1 The Derivative of Sin x

We are now going to use our knowledge of the trigonometric functions to evaluate their derivatives. Let us begin with $\sin x$. We wish to determine

$$\frac{d}{dx}(\sin x) = \lim_{h \to 0}\left(\frac{\sin(x + h) - \sin(x)}{h}\right)$$

From trigonometric identities we know that

$$\sin(x + h) = \sin(x)\cos(h) + \cos(x)\sin(h)$$

so that

$$\sin(x + h) - \sin(x) = \sin(x)\left[\cos(h) - 1\right] + \cos(x)\sin(h)$$

We can therefore write the derivative of $\sin x$ as

$$\frac{d}{dx}[\sin(x)] = \sin(x)\left[\lim_{h \to 0}\frac{\cos(h) - 1}{h}\right]$$

$$+ \cos(x)\left[\lim_{h \to 0}\frac{\sin(h)}{h}\right] \qquad (4.1)$$

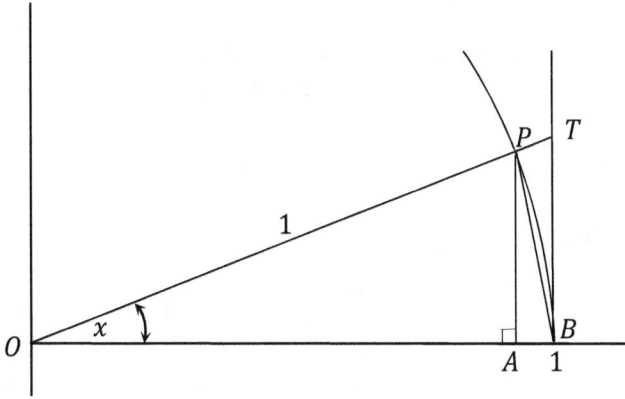

Fig. 4.1. Applying the squeeze theorem with the unit circle.

The problem now is to determine

$$\lim_{h \to 0} \left[\frac{\cos(h) - 1}{h} \right] \text{ and } \lim_{h \to 0} \frac{\sin (h)}{h}$$

The diagram in Fig. 4.1 shows part of the unit circle with several construction lines drawn. Consider triangle OPB and sector OPB. From triangle OPB we have

$$\text{Area } \Delta OPB = \frac{1}{2} \cdot OB \cdot AP$$

Now since our triangle is drawn within a unit circle then the hypotenuse is of length one and thus $AP = \sin x$. Also $OB = 1$, therefore

$$\text{Area } \Delta OPB = \frac{1}{2} \sin x$$

Now,

$$\text{Area sector } OPB = \frac{1}{2} r^2 x = \frac{1}{2} x$$

**

Notice that the area of the triangle will always be less than the sector, no matter how small angle x. Thus,

$$\frac{1}{2}\sin x < \frac{1}{2}x$$

Therefore,

$$\sin x < x$$

i.e.

$$\frac{\sin x}{x} < 1 \qquad (4.2)$$

Now consider the larger triangle OTB. The area of this triangle is greater than the sector OPB. That is,

$$\frac{1}{2}\cdot 1 \cdot \tan x > \frac{1}{2}x$$

$$\tan x > x$$

$$\frac{\sin x}{\cos x} > x$$

$$\frac{\sin x}{x} > \cos x$$

Now, as $x \to 0$, $\cos x \to 1$, therefore

$$\frac{\sin x}{x} > 1 \qquad (4.3)$$

By considering (4.2) and (4.3) we see that $\frac{\sin x}{x}$ gets 'squeezed' to 1 as $x \to 0$. That is,

---

** To find the area of a sector formula: equate the ratio of angle x to a full circle angle 2π, with the ratio of the area of the sector to the area of a circle.

$$\lim_{h \to 0} \frac{\sin h}{h} = 1$$

We now wish to find

$$\lim_{h \to 0} \frac{\cos h - 1}{h}$$

First, we multiply by the conjugate of the numerator. That is,

$$\lim_{h \to 0} \frac{\cos h - 1}{h} \cdot \frac{\cos h + 1}{\cos h + 1} = \lim_{h \to 0} \frac{\cos^2 h - 1}{h(\cos h + 1)}$$

$$= \lim_{h \to 0} \frac{\cos^2 h - (\sin^2 h + \cos^2 h)}{h(\cos h + 1)}$$

$$= \lim_{h \to 0} \frac{-\sin^2 h}{h(\cos h + 1)}$$

$$= \lim_{h \to 0} \frac{-\sin h}{h} \cdot \frac{\sin h}{\cos h + 1}$$

$$= -1 \cdot \frac{0}{1 + 1} = 0$$

That is,

$$\boxed{\lim_{h \to 0} \frac{\cos h - 1}{h} = 0}$$

We can now see from equation (4.1) that

$$\boxed{\frac{d}{dx}(\sin x) = \cos x}$$

Example 4.1. *Differentiate* sin 2x.

Here we have an example of a composite trigonometric function. We let

$$u(x) = \sin x, \qquad v(x) = 2x$$

so that utilising the chain rule we find,

$$\frac{d}{dx}(\sin 2x) = u'\big(v(x)\big)v'(x) = 2\cos 2x$$

$\square$

4.2 The Derivative of Cos x

We would like to now determine the derivative of $\cos x$. We have

$$\frac{d}{dx}[\cos x] = \lim_{h \to 0}\left[\frac{\cos(x + h) - \cos(x)}{h}\right]$$

Now, using the trigonometric identity

$$\cos(x + h) = \cos x \cos h - \sin x \sin h$$

we have

$$\frac{d}{dx}[\cos x] = \lim_{h \to 0}\frac{\cos x \cos h - \sin x \sin h - \cos x}{h}$$

$$= \lim_{h \to 0}\frac{\cos x\,[\cos h - 1] - \sin x \sin h}{h}$$

$$= \cos x \lim_{h \to 0}\left[\frac{\cos h - 1}{h}\right] - \sin x \lim_{h \to 0}\left[\frac{\sin h}{h}\right]$$

We know that

$$\lim_{h \to 0}\left[\frac{\cos(h) - 1}{h}\right] = 0$$

and

$$\lim_{h \to 0}\frac{\sin(h)}{h} = 1$$

We therefore see that

$$\frac{d}{dx}[\cos(x)] = \cos(x) \cdot 0 - \sin(x) \cdot 1$$

i.e.

$$\frac{d}{dx}(\cos x) = -\sin x$$

Example 4.2. *Find the derivative of* $y = \cos \frac{1}{2}x$.

As in the previous example we again apply the chain rule. So,

$$\frac{dy}{dx} = \frac{1}{2} \cdot \frac{d}{dx}\left(\cos\frac{1}{2}x\right) = -\frac{1}{2}\sin\frac{1}{2}x$$

□

4.3 The Derivative of Tan x

Knowing the derivatives of $\sin x$ and $\cos x$, we can now use our differentiation rules to form the derivatives of the other trigonometric functions quite easily. For example,

$$\tan x = \frac{\sin x}{\cos x}$$

so that

$$\frac{d}{dx}(\tan x) = \frac{d}{dx}\left(\frac{\sin x}{\cos x}\right) = \frac{\cos x \cdot \cos x - \sin x \cdot -\sin x}{\cos^2 x}$$

$$= \frac{\cos^2 x + \sin^2 x}{\cos^2 x} = \frac{1}{\cos^2 x}$$

We therefore see that

$$\frac{d}{dx}(\tan x) = \sec^2 x$$

Example 4.3. *Find the derivative of* $\tan^3 \theta$.

Solution:

$$\frac{d}{dx}(\tan^3 \theta) = 3(\tan \theta)^2 \cdot \sec^2 \theta$$

$$= 3 \tan^2 \theta \sec^2 \theta$$

□

Let us now develop formulas for finding the derivative of the reciprocal trigonometric functions.

4.4 The Derivative of the Reciprocal Trigonometric Functions

To find the derivative of $\sec x$ first write

$$\sec x = \frac{1}{\cos x} = (\cos x)^{-1}$$

so that

$$\frac{d}{dx}(\sec x) = \frac{d}{dx}((\cos x)^{-1})$$

$$= -(\cos x)^{-2} \cdot - \sin x$$

$$= \frac{\sin x}{\cos^2 x}$$

$$= \frac{\sin x}{\cos x} \cdot \frac{1}{\cos x}$$

We therefore see that

$$\boxed{\frac{d}{dx}(\sec x) = \sec x \tan x}$$

To find the derivative of $\operatorname{cosec} x$ (or $\csc x$) consider

$$\csc x = \frac{1}{\sin x}$$

so that

$$\frac{d}{dx}(\csc x) = \frac{d}{dx}\left(\frac{1}{\sin x}\right)$$

$$= \frac{d}{dx}((\sin x)^{-1})$$

$$= -(\sin x)^{-2} \cdot \cos x$$

i.e.

$$\frac{d}{dx}(\csc x) = -\frac{\cos x}{\sin^2 x} = -\frac{\cot x}{\sin x}$$

We therefore see that

$$\boxed{\frac{d}{dx}(\csc x) = -\csc x \cot x}$$

For the derivative of $\cot x$ we have

$$\cot x = \frac{1}{\tan x} = (\tan x)^{-1}$$

$$\therefore \frac{d}{dx}(\cot x) = -(\tan x)^{-2} \sec^2 x$$

$$= -\frac{1}{\tan^2 x} \cdot \frac{1}{\cos^2 x}$$

$$= -\frac{\cos^2 x}{\sin^2 x} \cdot \frac{1}{\cos^2 x}$$

$$= -\frac{1}{\sin^2 x}$$

i.e.

$$\boxed{\frac{d}{dx}(\cot x) = -\csc^2 x}$$

Exercise 4

1. Differentiate the following functions with respect to x.

i) $\cos^2 3x$ ii) $\cos(2x + c)$ iii) $\cos^2 x + \sin^2 x$

iv) $3 \sin x \cos x$ v) $\dfrac{x^2}{\sin x}$ vi) $\sqrt{\cos x}$

vii) $3x \sin^2 x$ viii) $2x^3 \tan x$ ix) $\cos 3x \sin 2x$

x) $x^2 \cos^3 x \sin x$ xi) $\dfrac{\cos x}{\sin x - 1}$ xii) $\dfrac{1 + \sin 3x}{1 - \sin 3x}$

xiii) $\dfrac{1}{\csc x - \tan x}$ xiv) $\tan^4 3x$ xv) $\cos^2(5x - 3)$

xvi) $\sqrt{\sec x} - \sin \sqrt{x}$ xvii) $\cos^3 2x \sin^2 3x$

xviii) $\dfrac{\cos 2x}{1 - \sin x}$ xix) $x^3(x^2 - \cos^2 x)^4$

xx) $\cot 3x + \sqrt[3]{\tan 2x}$ xxi) $\sin^2 2x \tan x$

xxii) $10x^5 \sec 2x$ xxiii) $\sec^2 3x \tan^4 3x$

xxiv) $\sin\left(\dfrac{x}{x^2+1}\right)$

2. For $y = a\sin \omega x + b\cos \omega x$ show that $\dfrac{d^2y}{dx^2} = -\omega^2 y$.

3. Find the second derivative of $A = \sqrt{\cos \theta} - \sin \theta$.

4. Find the equation of the tangent line to the curve

$$y = 2 \sin x - \frac{1}{\sin x}$$

where $x = \dfrac{\pi}{3}$.

5. Find the equation of the tangent line to the curve of the function $f(x) = x^2 \cos x$, at the point $(\pi, -\pi^2)$.

6. For what values of x does the curve of $y = \sqrt{3}x - \cos 2x$ have a horizontal gradient, given the domain $[0, 2\pi]$.

7. A particle's motion is defined by the equation $x = t + \cos 2t$. Find

a) the initial displacement, velocity and acceleration,

b) the velocity and acceleration when $t = \dfrac{\pi}{6}$,

   c) the amount of time between its first and second moments of rest.

8. The displacement of a body in rectilinear motion from an origin is given by $x = 20t \sin 2t$ cm after t seconds. What is the body's velocity and acceleration at $t = \frac{\pi}{3}$ s?

9. Given

$$f'(x) = \lim_{h \to 0} \frac{\frac{1}{\sin^2(x+h)} - \frac{1}{\sin^2 x}}{h}$$

determine $f'(\frac{\pi}{6})$.

10. Find $\frac{dw}{dx}$ given $w = 2y^{\frac{3}{2}}$ and $y = \cot x$.

11. Given $y = \csc^2 \theta$ show that $y'' = 2y^2(2\cos^2 \theta + 1)$.

12. i) Given $y = -2\cos \theta$ and $x = \tan \theta$ show that

$$\frac{dy}{dx} = \sin 2\theta \cos \theta$$

   ii) For $y = \frac{1}{2}\tan^2 \theta$ show that

$$\frac{d^2y}{d\theta^2} = (6y + 1)(2y + 1)$$

13. a) What does $\frac{d\theta}{d(\cos \theta)}$ equal?

   b) Find $\frac{d(\cos x)}{du}$ given $u = \tan x$.

14. Find the gradient of the tangent line to the curve

$$H = \frac{4\cos \theta + 3}{(\sin \theta - 1)^2}$$

at the point $\left(\frac{\pi}{3}, \frac{20}{7-4\sqrt{3}}\right)$.

Answers

1) i) $-3\sin 6x$ ii) $-2\sin(2x + c)$ iii) 0

iv) $3\cos 2x$ v) $\dfrac{x(2\sin x - x\cos x)}{\sin^2 x}$ vi) $-\dfrac{\sin x}{2\sqrt{\cos x}}$

vii) $3(x\sin 2x + \sin^2 x)$ viii) $\dfrac{x^2(2x + 3\sin 2x)}{\cos^2 x}$

ix) $\dfrac{1}{2}(5\cos 5x - \cos x)$

x) $x\cos^2 x\,[x\cos^2 x + \sin x\,(2\cos x - 3x)]$

xi) $\dfrac{2\sin x + 1}{\sin x - 1}$ xii) $\dfrac{-3\sin 6x}{(1 - \sin 3x)^2}$ xiii) $\dfrac{\cot x \cdot \csc x}{(\csc x - \tan x)^2}$

xiv) $12\tan^3 3x\sec^2 3x$ xv) $-5\sin[2(5x - 3)]$

xvi) $\dfrac{1}{2}\sqrt{\sec x}\tan x$ xvii) $6\cos^2 2x\sin 3x\cos 5x$

xviii) $\dfrac{\sin 2x(\sin x - 2) + \cos x}{(1 - \sin x)^2}$

xix) $x^2(x^2 - \cos^2 x)^3(11x^2 - 3\cos^2 x + 4x\sin 2x)$

xx) $\dfrac{2}{3}\sqrt[3]{\csc^2 2x\sec^4 2x} - 3\csc^2 3x$

xxi) $\sin^2 2x\sec^2 x + 2\sin 4x\tan x$

xxii) $10x^4\sec 2x\,(2x\tan 2x + 5)$

xxiii) $6\sec^4 3x\tan^3 3x\,(\sin^2 3x + 2)$

xxiv) $\left[\dfrac{1 - x^2}{(x^2 + 1)^2}\right]\cos\left(\dfrac{x}{x^2 + 1}\right)$

2) $-\omega^2 y$

3) $\dfrac{3 - \sin 2\theta}{4\sqrt{(\cos\theta - \sin\theta)^3}}$

4) $3y - 5x = \sqrt{3}$

5) $2\pi x + y = \pi^2$

6) $x = \dfrac{2\pi}{3}, \dfrac{5\pi}{6}, \dfrac{5\pi}{3}, \dfrac{11\pi}{6}$

7) a) 1, 1, -4

   b) $1 - \sqrt{3}, -2$ c) $\pi/3$

8) $\dfrac{10}{3}(3\sqrt{3} - \pi)$ cm/s, $-\dfrac{40}{3}(3 + \pi\sqrt{3})$ cm/s^2

9) $-8\sqrt{3}$

10) $3\sqrt{\cot x} - \csc^2 x$

11) $2y^2(2\cos^2\theta + 1)$

13) a) $-\csc\theta$ b) $-\sin x\cos^2 x$

14) $\dfrac{2(2\sqrt{3} + 1)}{2 + \sqrt{3}}$

Chapter 5

Implicit Differentiation

5.1 Finding the Derivative of Implicit Equations

We have learnt in previous chapters how to find the derivative of various function types. These have all been *explicit* functions, that is, those where the independent variable (typically 'x') expression, is written separately from the dependent variable ('y'). However, there are some equations in which the 'x' expression cannot be written explicitly from 'y'; we say it is *implicit*.

The type of equations we are considering may involve either a *function* or *relation*. A function maps an x-value to a maximum of one y-value whereas a relation can map to multiple y-values. A geometrical interpretation of a function finds a vertical line intersecting once only for any part of the associated curve. Whereas, for a curve of a relation, a vertical line will intersect it more than once. See Fig. 5.1 below for an illustration of this.

For our purposes the difference between a function and a relation is not particularly important. Suffice to say that many instances requiring the use of *implicit differentiation* involves relations rather than functions.

Some examples of implicit equations are

$$x^2 - 2xy + 3y^2 - 7 = 0, \quad 5x^3y^2 + {}^1\!/_y = 4, \quad \sin(xy) = 2y$$

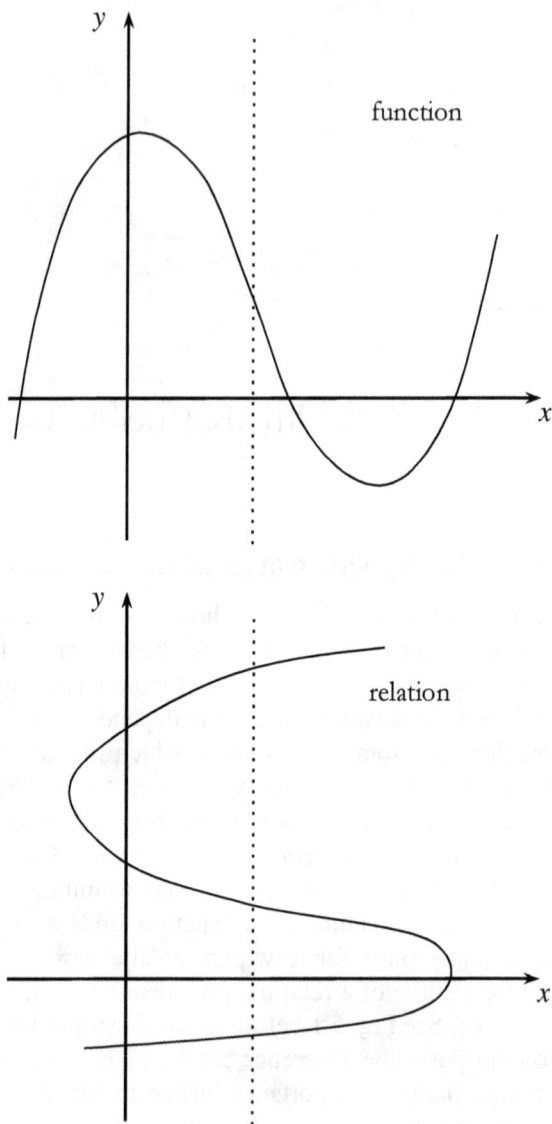

Fig. 5. 1. Geometrical interpretation of functions and relations.

The first equation here could be written explicitly in terms of x by utilising the *quadratic formula*. However, this is an unnecessary complication and finding the derivative is more easily achieved utilising implicit differentiation. To enable us to take the derivative of an implicit equation we make use of the chain rule. In an equation, if we represent a term that is a function of the dependent variable by $f(y)$ say, then differentiating with respect to x the chain rule says:

$$\frac{df}{dx} = \frac{df}{dy} \cdot \frac{dy}{dx}$$

For example, to differentiate the expression $3y^2$ with respect to x we have

$$\frac{d}{dx}(3y^2) = \frac{d}{dy}(3y^2) \cdot \frac{dy}{dx} = 6y \cdot \frac{dy}{dx}$$

An equation may involve more complex terms. Differentiating an expression involving a product for example, we would need to make use of both the chain and product rule. E.g.

$$\frac{d}{dx}(2xy^3) = 2x\frac{d}{dx}(y^3) + y^3\frac{d}{dx}(2x)$$

$$= 2x\frac{d}{dy}(y^3) \cdot \frac{dy}{dx} + y^3 \cdot 2$$

$$= 2x \cdot 3y^2\frac{dy}{dx} + 2y^3$$

$$= 6xy^2\frac{dy}{dx} + 2y^3$$

Example 5.1 *Find* $\frac{dy}{dx}$ *from the equation* $x^2 + y^2 - 3x + 5y = 9$.

Here, it is difficult to express y as an explicit function of x, so we will apply implicit differentiation to the equation instead. That is,

$$2x + 2y\frac{dy}{dx} - 3 + 5\frac{dy}{dx} = 0$$

$$\frac{dy}{dx}(2y + 5) = 3 - 2x$$

$$\frac{dy}{dx} = \frac{3 - 2x}{2y + 5}$$

☐

Example 5.2. *Use implicit differentiation to find $\frac{dy}{dx}$ for the following equation:*

$$4x \sin y - 2y^3 \cos x + 7 = 0$$

Differentiate with respect to x:

$$4x \cos y \frac{dy}{dx} + 4 \sin y + 2y^3 \sin x - 6y^2 \frac{dy}{dx} \cos x = 0$$

$$4x \cos y \frac{dy}{dx} - 6y^2 \cos x \frac{dy}{dx} = -4 \sin y - 2y^3 \sin x$$

$$\frac{dy}{dx}(4x \cos y - 6y^2 \cos x) = -2(2 \sin y + y^3 \sin x)$$

$$\frac{dy}{dx} = \frac{-2(2 \sin y + y^3 \sin x)}{4x \cos y - 6y^2 \cos x} = \frac{2 \sin y + y^3 \sin x}{3y^2 \cos x - 2x \cos y}$$

☐

Example 5.3. *Find the derivative function for the equation*

$$\frac{(5 - 2y^4)^2}{4x^2 - 3} + y = 10$$

Differentiate with respect to x:

$$\frac{(4x^2 - 3) \cdot 2(5 - 2y^4)\left(-8y^3 \frac{dy}{dx}\right) - (5 - 2y^4)^2 \cdot 8x}{(4x^2 - 3)^2} + \frac{dy}{dx}$$

$$= 0$$

$$-16y^3(4x^2 - 3)(5 - 2y^4)\frac{dy}{dx} - 8x(5 - 2y^4)^2$$

$$+ (4x^2 - 3)^2\frac{dy}{dx} = 0$$

$$\therefore \frac{dy}{dx} = \frac{8x(5 - 2y^4)^2}{(4x^2 - 3)^2 - 16y^3(4x^2 - 3)(5 - 2y^4)}$$

$$= \frac{8x(5 - 2y^4)^2}{(4x^2 - 3)[4x^2 - 3 - 16y^3(5 - 2y^4)]}$$

□

Example 5.4. *Find the second derivative for the relation described by* $2x^2 - 3y^2 - 4x + 6y + 11 = 0$ *and determine its value when* $x = 1$ *and* $y = 2$.

Differentiating with respect to x we find,

$$4x - 6y\frac{dy}{dx} - 4 + 6\frac{dy}{dx} = 0$$

$$(6 - 6y)\frac{dy}{dx} = 4 - 4x$$

$$\frac{dy}{dx} = \frac{4 - 4x}{6 - 6y} = \frac{2 - 2x}{3 - 3y} = \frac{2}{3}\frac{1 - x}{1 - y} \qquad (5.1)$$

Differentiating again yields,

$$\frac{d^2y}{dx^2} = \frac{2}{3} \cdot \frac{(1 - y)(-1) - (1 - x)(-1)\frac{dy}{dx}}{(1 - y)^2}$$

$$= \frac{2}{3} \cdot \frac{(y - 1) + (1 - x)\frac{dy}{dx}}{(1 - y)^2}$$

Now, substituting for dy/dx from (5.1) gives,

$$\frac{d^2y}{dx^2} = \frac{2}{3} \cdot \frac{y - 1 + (1 - x)\cdot\frac{2}{3}\cdot\frac{1-x}{1-y}}{(1 - y)^2}$$

$$= \frac{2}{3} \cdot \frac{\dfrac{-(y-1)^2 + \frac{2}{3}(1-x)^2}{1-y}}{(1-y)^2}$$

$$= \frac{2}{3} \cdot \frac{\frac{2}{3}(1-x)^2 - (y-1)^2}{(1-y)^3}$$

$$\therefore \left. \frac{d^2 y}{dx^2} \right|_{x=1, y=2} = \frac{2}{3} \cdot \frac{\frac{2}{3}(1-1)^2 - (2-1)^2}{(1-2)^3}$$

$$= \frac{2}{3} \cdot \frac{-1}{-1} = \frac{2}{3}$$

□

5.2 The Derivative of Inverse Trigonometric Functions

We seek to find the derivative of the inverse trigonometric functions $\sin^{-1} x$, $\cos^{-1} x$ and $\tan^{-1} x$.[††] We can achieve this by utilising implicit differentiation. Firstly, in finding the derivative of $\sin^{-1} x$ we let

$$x = \sin y \qquad\qquad -1 \le x \le 1$$

noting the usual restriction that applies to the sine function. Applying the derivative with respect to x to this equation gives

$$1 = \cos y \frac{dy}{dx}$$

$$\therefore \frac{dy}{dx} = \frac{1}{\cos y} \qquad\qquad (5.2)$$

However, we require a derived function in terms of x, not y. We therefore express $\cos y$ in (5.2) in terms of x by considering

---

[††] Trigonometry has an unfortunate notation for inverse functions. Here we are representing the inverse, not the reciprocal, of the trigonometric functions. Note also an alternative notation is sometimes used: $\arcsin x$, $\arccos x$, $\arctan x$.

$$\sin y = x = \frac{x}{1}$$

in its relation within a right angled triangle. From Fig. (5.2) we see that the sine of angle y requires a side of length x and a hypotenuse of length 1. Utilising Pythagoras's theorem we can find the length of side b in terms of x. That is,

$$b = \sqrt{1 - x^2}$$

so that

$$\cos y = \frac{b}{1} = \sqrt{1 - x^2}$$

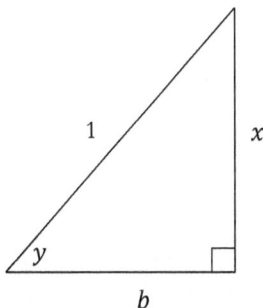

Fig. 5.2. Expressing $\cos y$ in terms of x.

Hence, (5.2) can be written

$$\frac{dy}{dx} = \frac{1}{\sqrt{1 - x^2}}$$

That is,

$$\frac{d}{dx}(\sin^{-1} x) = \frac{1}{\sqrt{1 - x^2}}, \qquad -1 < x < 1$$

Note the further restriction on x; we cannot allow $x = \pm 1$ as otherwise an infinity would result. Thus the final domain becomes $x \in (-1, 1)$.

Adopting the same method for $\cos^{-1} x$ we let

$$x = \cos y \qquad\qquad -1 \le x \le 1$$

Taking the derivative with respect to x gives

$$1 = -\sin y \cdot \frac{dy}{dx}$$

$$\therefore \frac{dy}{dx} = -\frac{1}{\sin y}$$

From Fig. 5.3 we see that

$$\sin y = \frac{\sqrt{1 - x^2}}{1} = \sqrt{1 - x^2}$$

So,

$$\boxed{\frac{d}{dx}(\cos^{-1} x) = -\frac{1}{\sqrt{1 - x^2}}, \qquad -1 < x < 1}$$

In finding the derivative of $\tan^{-1} x$ we let

$$x = \tan y, \qquad x \in \mathbb{R}$$

Taking the derivative gives

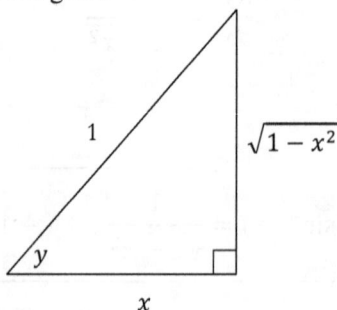

Fig. 5.3. Expressing $\sin y$ in terms of x.

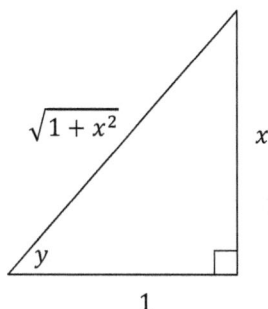

Fig. 5.4. Expressing tan y in terms of x.

$$1 = \sec^2 y \frac{dy}{dx}$$

$$\therefore \frac{dy}{dx} = \frac{1}{\sec^2 y} = \cos^2 y$$

and from Fig. 5.4

$$\cos y = \frac{1}{\sqrt{1 + x^2}}$$

$$\therefore \cos^2 y = \frac{1}{1 + x^2}$$

i.e.

$$\boxed{\frac{d}{dx}(\tan^{-1} x) = \frac{1}{1 + x^2}, \quad x \in \mathbb{R}}$$

Example 5.5. *Find the derivative of* $\sin^{-1} 5x$.

Let $f(x) = \sin^{-1} x$ and $g(x) = 5x$. Therefore,

$$f'(g(x)) = \frac{1}{\sqrt{1 - (5x)^2}} \cdot 5$$

by the chain rule. That is, the derived function is given by

$$\frac{5}{\sqrt{1 - 25x^2}}$$

☐

Example 5.6. *Differentiate* (a) $y = x \cos^{-1} \sqrt{x}$ (b) $y = \dfrac{1}{\tan^{-1} x}$

(a)

$$\frac{dy}{dx} = x \cdot \frac{d}{dx} \left(\cos^{-1} \sqrt{x} \right) + \cos^{-1} \sqrt{x} \cdot \frac{d}{dx} (x)$$

$$= x \cdot \frac{-1}{\sqrt{1 - \left(\sqrt{x}\right)^2}} \cdot \frac{1}{2} x^{-\frac{1}{2}} + \cos^{-1} \sqrt{x} \cdot 1$$

$$= \frac{-\sqrt{x}}{2\sqrt{1 - x}} + \cos^{-1} \sqrt{x}$$

$$= \frac{2\sqrt{1 - x} \cos^{-1} \sqrt{x} - \sqrt{x}}{2\sqrt{1 - x}}$$

(b)

$$\frac{dy}{dx} = -(\tan^{-1} x)^{-2} \cdot \frac{1}{x^2 + 1}$$

$$= -\frac{1}{(x^2 + 1)(\tan^{-1} x)^2}$$

☐

Exercise 5

1. For $x^3 + y^3 - 27xy = 1$, determine $\frac{dy}{dx}$ when $x = 1, y = 2$.

2. For $\sqrt{x} - \sqrt{y} = 9$, find $\left[\frac{dy}{dx}\right]_{x=16}$

3. For the differential equation

$$3y^2 \frac{d^2y}{dx^2} + 6y \left(\frac{dy}{dx}\right)^2 + y^3 = k$$

   find the value of k given that $y^3 = 2 - \sin x$ is a solution.

4. For the curve described by $y^3 - 2xy + 4x^2 = -5$, find the value of $\frac{d^2y}{dx^2}$ at the point $(-1, 1)$.

5. Find the equation of the tangent line to the circle defined by $(x - 3)^2 + (y - 1)^2 = 16$, at the point $\left(4, \sqrt{7} + 1\right)$.

6. Find dy/dx by implicit differentiation for the following equations.

   i. $y \tan(2x - y) = 6$

   ii. $\cos^2(x - y) + \sin^2(3x + y) = \cos(x + 2y)$

   iii. $(x^2 + y^2)^2 = 5x^3y^4$

7. Find the equation of the tangent line to the curve described by $\cos(y) = x$, where the tangent and the curve intersect at the point $\left(\frac{\sqrt{3}}{2}, \frac{\pi}{6}\right)$.

8. Find the derivative of the following functions.

   (i) $y = (\cos^{-1} x)^3$ (ii) $f(x) = \sqrt{1 - x^2} \sin^{-1} x$

   (iii) $h(x) = \tan^{-1}(e^{2x})$ (iv) $g(x) = (1 - x^2)\ln(\sin^{-1} x)$

   (v) $x(t) = e^{\cos^{-1} t}$ (vi) $y = \cos^{-1}(\sin \theta)$

   (vii) $A(x) = \tan^{-1}[\sqrt{x + 1}(x^2 + 1)]$

   (viii) $B(x) = \sin^{-1}\left(\frac{\tan x}{x^2 + 1}\right)$

9. Find the gradient functions and their domains of the following functions.

   (a) $g(x) = \cos^{-1}(e^x)$ (b) $h(x) = \tan^{-1}\left(\frac{2}{x}\right) + \sqrt{1 - \frac{4}{x^2}}$

   (c) $\theta(x) = \sin^{-1}[\tan(\cos^{-1} x)]$ (d) $f(x) = \frac{\tan^{-1} x}{\sin^{-1} x}$

10. If $g(x) = x^2 \arccos(x/2) - \sqrt{9 - x^2}$, find the value of $g'(\sqrt{3})$, $0 < x < \pi$.

11. For the function $f(x) = \ln[\tan^{-1}(5x + 3)]$, $\frac{6\pi - 3}{5} < x < \frac{13\pi - 6}{10}$ find $f'(\pi)$.

12. Find the equation of the tangent line to the curve $4 \arcsin\frac{2}{x}$ at the point $\left(e, \frac{1}{2}\right)$.

13. Find $\frac{d^2 y}{dx^2}$ for $y = -2\sin(\arctan(3\sqrt{x}))$.

14. Take the derivative with respect to x of

$$x^3 - 2y^3 = \cos y$$

15. What is the value of the second derivative on the curve described by

$$2y^3 + x^3 y - 4x^2 = 5$$

   at the point $(\sqrt{2}, -1)$?

16. Find the equation(s) of the tangent(s) to the curve

$$y^2 \sin x + y = 2x$$

   at the point $\left(\frac{\pi}{2}, \frac{1 \pm \sqrt{4\pi + 1}}{2}\right)$.

17. Determine the equation(s) of any horizontal and vertical tangent(s) to the curve described by $2x^2 - 9x + y^3 - 3y^2 = -9$.

Answers

1) $-\dfrac{17}{5}$

2) $\dfrac{5}{4}$

3) 2

4) $-\dfrac{12}{5}$

5) $x + \sqrt{7}y = 11 + \sqrt{7}$

6) a) $\dfrac{4y + \sin(4x - 2y)}{2y}$ b) $\dfrac{\sin[2(x-y)] - 3\sin[2(3x+y)] - \sin(x+2y)}{2\{\sin[2(x-y)] + \sin(x+2y)\}}$

c) $\dfrac{15x^2y^4 - x^3 - xy^2}{y(x^2 + y^2 - 5x^3y^2)}$

7) $2x + y = \dfrac{\pi}{6} + \sqrt{3}$

8) i) $-\dfrac{3(\cos^{-1}x)^2}{\sqrt{1-x^2}}$ ii) $1 + \dfrac{x\sin^{-1}x}{\sqrt{1-x^2}}$ iii) $\dfrac{2e^{2x}}{1+e^{4x}}$

iv) $\sqrt{1-x^2}\csc^{-1}x - 2x\ln(\sin^{-1}x)$ v) $-\dfrac{1}{\sqrt{1-t^2}}e^{\cos^{-1}t}$

vi) -1 vii) $\dfrac{3x+1}{2\sqrt{x+1}(x^2+1)^2}$ viii) $\dfrac{(x^2+1)\sec^2 x - 2x\tan x}{(x^2+1)\sqrt{(x^2+1)^2 - \tan^2 x}}$

9) a) $-\dfrac{e^x}{\sqrt{1-e^{2x}}}$; $x \le 0$

b) $\dfrac{2[2(x^2+4) - x^2\sqrt{x^2-4}]}{x^2(x^2+4)\sqrt{x^2-4}}$; $x > 2 \cup x < -2$

c) $-\dfrac{\sec^2(\cos^{-1}x)}{\sqrt{(1-x^2)(1-\tan^2(\cos^{-1}x))}}$; $\dfrac{1}{\sqrt{2}} < x < 1$

d) $\dfrac{\sin^{-1}x - \sqrt{1-x^2}\tan^{-1}x}{(1-x^2)(\sin^{-1}x)^2}$; $-1 < x < 0 \cap 0 < x < 1$

10) $\dfrac{2\pi + 3(\sqrt{2} - 6)}{6}$ 11) 1.64×10^{-4}

12) $8x + 2e\sqrt{e^2 - 4}y = e(16 + \sqrt{e^2 - 4})$

13) $\dfrac{27(2+\sqrt{2})}{2\sqrt{x}(9x+1)^{\frac{5}{2}}}$

14) $\dfrac{3x^2}{6y^2 - \sin y}$

15) $\dfrac{159\sqrt{2} + 416}{2(\sqrt{2}+3)^3}$

16) $2(2 \pm \sqrt{4\pi + 1})y = 4x + 3 \pm 3\sqrt{4\pi + 1} + 4\pi$

17) $x = \dfrac{3}{2}, x = 3, x = \dfrac{9 \pm \sqrt{41}}{4}, y \approx 3.12$

Chapter 6

The Derivative of Exponential & Logarithmic Functions

6.1 The Derivative of Exponential Functions

We seek to find the derivative of an exponential function such as for example, $f(x) = a^x$. Finding the derivative by first principles we have

$$f'(x) = \lim_{h \to 0} \frac{a^{x+h} - a^x}{h}$$

$$= \lim_{h \to 0} \frac{a^x(a^h - 1)}{h}$$

$$= a^x \lim_{h \to 0} \frac{a^h - 1}{h} \qquad (6.1)$$

Now it turns out that when a is a special number known as *Euler's number*, $a = 2.71828...$, that the limit in (6.1) equals 1. Therefore, $f'(x) = a^x$. Mathematicians have given Euler's number a designation of e and we have found that for e^x, its derivative is itself! That is, the gradient at any point of the curve e^x is the value of the function at that point.

For a more complicated function such as $f(x) = e^{g(x)}$, we make use of the chain rule to determine its derivative. For example, for some function $y = f[g(x)]$ where $f(x) = e^x$, we find $y' = f'[g(x)]g'(x)$. Since $f'(x) = e^x$ we then have

$$\boxed{y' = g'(x)e^{g(x)}}$$

$$(6.2)$$

Example 6.1. *Differentiate the function* $y = e^{3x^2 - 5}$.

The derivative of the index for this exponential function is $6x$. Therefore the derivative of the function given in the example is simply $6x$ multiplied by the original function. i.e.

$$\frac{dy}{dx} = 6xe^{3x^2 - 5}$$

□

Example 6.2. *Find* dy/dx *of* $y = e^{2\tan x}$.

Taking the derivative gives

$$\frac{dy}{dx} = 2\sec^2 x \cdot e^{2\tan x}$$

□

To find the derivative of an exponential function that has a different base we first need to understand how to take the derivative of a logarithmic function. We will explore this in the following section.

6.2 The Derivative of Logarithmic Functions

We will begin by finding the derivative of a *log* function with base e. Given some function f defined by the [‡‡]*natural logarithm*

[‡] A natural logarithm is a logarithm to the base of e. i.e.
$\ln g(x) \equiv \log_e g(x)$.

$$f(x) = \ln g(x) \qquad (6.3)$$

we seek to find its derivative. First note that (6.3) can be written as the equivalent equation

$$g(x) = e^{f(x)} \qquad (6.4)$$

Now, differentiating (6.4) as per the previous section, we find

$$g'(x) = f'(x)e^{f(x)}$$

Rearranging to make f' the subject,

$$f'(x) = \frac{g'(x)}{e^{f(x)}} \qquad (6.5)$$

And substituting (6.4) we have

$$f'(x) = \frac{g'(x)}{g(x)}$$

i.e.

$$\boxed{\frac{d}{dx}(\ln g(x)) = \frac{g'(x)}{g(x)}}$$

That is, we have found that the derivative of a natural logarithm function is simply the argument of the *ln* function on the denominator of a fraction, with the derivative of the argument on the numerator.

Example 6.3 *Find the derivative of the function* $y = \ln \sqrt{x}$.

$$\frac{dy}{dx} = \frac{\frac{1}{2}x^{-\frac{1}{2}}}{x^{\frac{1}{2}}} = \frac{1}{2x^{\frac{1}{2}+\frac{1}{2}}} = \frac{1}{2x}$$

$\square$

Example 6.4. *Find the derivative of* $g(x) = \ln \sqrt{9 - x^2}$.

First, convert the surd into fractional indicial form, i.e.

$$g(x) = \ln(9 - x^2)^{\frac{1}{2}}$$

Now,

$$g'(x) = \frac{\frac{1}{2}(9-x^2)^{-\frac{1}{2}}(-2x)}{(9 - x^2)^{\frac{1}{2}}}$$

$$= \frac{-x}{(9 - x^2)^{\frac{1}{2}+\frac{1}{2}}} = \frac{x}{x^2 - 9}$$

□

Example 6.5. *What is the gradient function for* $f(x) = \ln(\sin^2 3x)$?

We have,

$$f'(x) = \frac{2[\sin 3x][3\cos 3x]}{\sin^2 3x}$$

$$= \frac{6\cos 3x}{\sin 3x} = 6\cot 3x$$

□

Let us now find the derivative of a log function with any base a. To achieve this we need to apply the change of base rule for logarithms, i.e.

$$\log_a x = \frac{\log_c x}{\log_c a}$$

where a and c are constants. However, in replacement of c we will use Euler's number e to form a natural logarithm. So,

$$\log_a f(x) = \frac{\log_e f(x)}{\log_e a} = \frac{\ln f(x)}{\ln a}$$

Taking the derivative and treating $\frac{1}{\ln a}$ as a coefficient we find

$$\frac{d}{dx}(\log_a f(x)) = \frac{1}{\ln a} \cdot \frac{d}{dx}(\ln f(x)) = \frac{1}{\ln a} \cdot \frac{f'(x)}{f(x)}$$

Example 6.6. *Find the derivative of* $\log \sqrt[3]{6x^2 - 7}$.

Let

$$y = \log\left[(6x^2 - 7)^{\frac{1}{3}}\right]$$

So,

$$\frac{dy}{dx} = \frac{1}{\ln 10} \cdot \frac{\frac{1}{3}(6x^2 - 7)^{-\frac{2}{3}} \cdot 12x}{(6x^2 - 7)^{\frac{1}{3}}}$$

$$= \frac{1}{\ln 10} \cdot \frac{4x}{(6x^2 - 7)^{\frac{1}{2} - (-\frac{2}{3})}}$$

$$= \frac{1}{\ln 10} \cdot \frac{4x}{6x^2 - 7}$$

□

Example 6.7. *Determine the gradient function of* $f(x) = \log_2[\sin(x^3)]$.

We have

$$f'(x) = \frac{1}{\ln 2} \cdot \frac{3x^2 \cos x^3}{\sin x^3}$$

$$= \frac{3x^2 \cot x^3}{\ln 2}$$

□

6.3 Logarithmic Differentiation

There are some equations where the derivative cannot be found directly and the form of the equation needs to be altered first. For example, the derivative of the expression x^x cannot be found in its current form so we alter it by applying logarithms. To achieve logarithmic differentiation we simply apply the natural logarithm to an equation followed by implicit differentiation. As an example, in differentiating $y = a^x$ (a constant) we first apply the natural logarithm to each side of the equation and then take the derivative. That is,

$$\ln y = \ln a^x$$

or,

$$\ln y = x \ln a$$

So,

$$\frac{1}{y} \cdot \frac{dy}{dx} = \ln a$$

i.e.

$$\frac{dy}{dx} = y \ln a$$

$$= a^x \ln a$$

Example 6.8. Differentiate $2^{\sqrt[3]{x^2}}$.

Let

$$y = 2^{\sqrt[3]{x^2}} = 2^{x^{\frac{2}{3}}}$$

Applying ln gives,

$$\ln y = \ln 2^{x^{\frac{2}{3}}} = x^{\frac{2}{3}} \cdot \ln 2$$

Differentiating,

$$\frac{1}{y} \cdot \frac{dy}{dx} = \frac{2}{3}x^{-\frac{1}{3}} \cdot \ln 2$$

$$\therefore \frac{dy}{dx} = \frac{2}{3} \cdot \frac{y}{x^{\frac{1}{3}}} \cdot \ln 2$$

$$= \frac{2}{3} \ln 2 \cdot \frac{2^{\sqrt[3]{x^2}}}{\sqrt[3]{x}}$$

□

Example 6.9. *Find the derivative of* $y = x^{\ln(2x)}$.

Applying the natural logarithm:

$$\ln y = \ln x^{\ln(2x)} = \ln(2x) \cdot \ln x$$

$$\therefore \frac{1}{y}\frac{dy}{dx} = \ln 2x \cdot \frac{1}{x} + \ln x \cdot \frac{2}{2x}$$

$$\frac{dy}{dx} = x^{\ln 2x}\left(\frac{\ln 2x + \ln x}{x}\right)$$

$$= x^{\ln 2x}\left(\frac{\ln 2x^2}{x}\right)$$

$$= x^{\ln 2x - 1}\ln 2x^2$$

$\square$

Exercise 6

1. Find the derivative of the following expressions:

   (i) $\ln\left(\sqrt{x^2 + 4} + 4\right)$ (ii) $\ln(e^x + e^{-x})$ (iii) e^{x^x}

   (iv) $\log_2\left(\frac{x^2-9}{x^2+9}\right)$ (v) $\log\left(\sqrt{1 + x^2}\right)$

   (vi) $\log\left(\sqrt[3]{x+1} - \sqrt[3]{x-1}\right)$ (vii) $\ln(\ln x)$ (viii) $e^{\cos x}$

   (ix) $\sin(e^{x^2})$ (x) $e^{\frac{a}{x}}$ (xi) $2x^3 e^{5\ln x}$ (xii) $\sin x \cdot e^{\cos x}$

   (xiii) $\ln\left(\frac{e^x}{e^x+2}\right)$ (xiv) $(\ln x)(\ln \sqrt{1-x})$ (xv) $\log_e(3\cot^2 x)$

2. Use the technique of logarithmic differentiation to find the derivative of the following:

   (i) $\left(\frac{x}{a}\right)^{ax}$ (ii) x^{x^x} (iii) 2^{3x} (iv) $(\cos x)^{x^2}$

   (v) $(1 - 2x - x^2)^{\frac{1}{x}}$ (vi) $5x(\sin x)^{3x^2}$ (vii) $\sqrt{7x}^{\sqrt{x}}$

   (viii) $(\pi x + 2)^{\sin x}$

3. Find the equation of the tangent line at $x = 1$ for the following function:

   $$f(x) = \frac{x^4 \cdot 3^{\ln x}(5x - 3)}{e^{x^2}(x + 1)^2}$$

4. Determine the gradient function for the following equations, assuming x is the independent variable.

   a) $\ln(xy) + y - 2x = 6$

   b) $xy^2 - \ln\left(\frac{y}{x}\right) = 1$

   c) $(\ln(3xy))^2 - 5xy = 2$

   d) $\ln\sqrt{2xy} = \sqrt{xy^2}$

5. Differentiate the following functions by logarithmic differentiation and simplify.

a) $y = \sqrt{\dfrac{2x-3}{x^3+1}}$

b) $y = \dfrac{\sqrt[3]{4x^3-1}}{\sqrt[5]{x^6+27}}$

c) $y = \dfrac{(x^2-1)(x^3+1)}{(x-4)^2}$

d) $y = \dfrac{\ln(x^2+1)}{\ln(x^2-1)}$

Answers

1) i) $\dfrac{x}{\sqrt{x^2+4}\left(\sqrt{x^2+4}+4\right)}$ ii) $\dfrac{e^x-e^{-x}}{e^x+e^{-x}}$ iii) $2xe^{xx}$ iv) $\dfrac{36x}{\ln 2(x^4-81)}$ v) $\dfrac{x}{\ln 10(1+x^2)}$

vi) $\dfrac{1}{9\ln 10\sqrt[3]{(x^2-1)^2}\left(\sqrt[3]{x+1}-\sqrt[3]{x-1}\right)}$ vii) $\dfrac{1}{x\ln x}$ viii) $-\sin x\, e^{\cos x}$

ix) $2x\, e^{x^2}\cos(e^{x^2})$ x) $-\dfrac{a}{x^2}e^{\frac{a}{x}}$ xi) $16x^7$ xii) $e^{\cos x}(\cos x - \sin^2 x)$

xiii) $\dfrac{2}{e^x+2}$ xiv) $\dfrac{(1-x)\ln(1-x)-x\ln x}{2x(1-x)}$ xv) $-\dfrac{2\csc^2 x}{\cot x}$

2) i) $a\left(\dfrac{x}{a}\right)^{ax}\left(1+\ln\dfrac{x}{a}\right)$ ii) $\left(\dfrac{x}{a}\right)^{ax}\left(x^{x-1}+x^x(1+\ln x)\ln x\right)$

iii) $2^{3x}\ln 8$ iv) $(\cos x)^{x^2}\left(2x\ln\cos x - x^2\tan x\right)$

v) $-\dfrac{(1-2x-x^2)^{\frac{1}{x}}}{x}\left[\dfrac{2(x+1)}{1-2x-x^2}+\dfrac{1}{x}(1-2x-x^2)\right]$

vi) $5(\sin x)^{3x^2}(1+3x^3\cot x+6x^2\ln\sin x)$

vii) $\dfrac{\sqrt{7x}^{\sqrt{x}}}{4\sqrt{x}}(2+\ln 7x)$ viii) $(\pi x+2)^{\sin x}\left(\dfrac{\pi}{\pi x+2}+\ln(\pi x+2)^{\cos x}\right)$

3) $4ey-(7+2\ln 3)x = -2e(5+2\ln 3)$

4) a) $\dfrac{2-y}{x(y+1)}$ b) $\dfrac{y(xy^2+1)}{x(1-2xy^2)}$ c) $\dfrac{y[5x^2-2\ln(3xy)]}{x[2\ln(3xy)-5xy]}$ d) $\dfrac{\sqrt{x}y-1}{x^2(1-2\sqrt{x}y)}$

5) a) $\dfrac{(9x^2-4x^3+2)y}{2(2x-3)(x^3+1)}$ b) $\dfrac{2x^2(3x^3-10x^6+49)y}{5(4x^3-1)(x^6+27)}$ c) $\dfrac{6x^5-20x^4-4x^3+15x^2-8x-1}{(x-4)^3}$

d) $\dfrac{2x[(x^2-1)\ln(x^2-1)-(x^2+1)\ln(x^2+1)]}{[\ln(x^2-1)]^2[x^4-1]}$

Chapter 7

Further Applications
of Differentiation

7.1 Curve Sketching

Calculus techniques can be utilised to gain information about the curve of a function. Certain key points on a curve may be derived, plotted and a smooth curve drawn connecting them. Though with the advent of modern graphics calculators and computers the need for manual curve sketching isn't always necessary, it is nonetheless an important skill to have. For simple graphs it is often quicker to sketch curves manually and it also proves a useful exercise in gaining a deeper understanding of the behaviour of different function types.

We saw in chapter 1 how the Fermat-Newton technique was used to locate the turning point of a parabola. We can use the same process to locate the turning points of any function. The turning point is found by finding where the derivative equals zero. These points are called *stationary points* or *critical points.*

It is usually required to test the extremum found to determine whether it is a maximum or minimum. There are two tests that can determine the nature of turning points. One is called the sign test, the other, the *second derivative test.*

Considering the sign test first, we find the gradient immediately to the left and right of the turning point. A minimum turning point has its gradient range from negative to zero to positive as we move along the curve from left to right. A maximum has the gradient change from positive through to negative. This is illustrated in Fig. 7.1 below.

When using the sign test the question arises as to the appropriate numbers to substitute to obtain the gradient value. For example we may inadvertently choose x-values that lie outside the turning point region. Also, there are three calculations required. A better option therefore, may be the second derivative test.

Consider the behaviour of the gradient as we move through a turning point. For a *maximum* the gradient's rate of change is *negative*. That is, representing the situation using the Leibniz notation we have

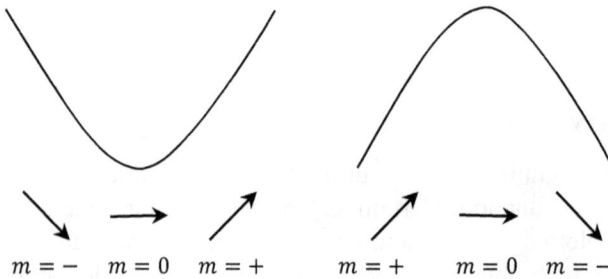

$$m = - \quad m = 0 \quad m = + \qquad m = + \quad m = 0 \quad m = -$$

Fig. 7. 1. Change in gradient around turning points.

$$\frac{dm}{dx} = \frac{d\left(\frac{dy}{dx}\right)}{dx} = \frac{d^2 y}{dx^2} < 0$$

where m is the gradient of a curve. For a *minimum* turning point the rate of change of the gradient is *positive*. That is,

$$\frac{d^2 y}{dx^2} > 0$$

With this test we simply take the second derivative and substitute in the x-value of the turning point. Note that depending on the complexity of the gradient function, determining the second derivative may require a fair amount of working and the sign test would then become more appropriate. It should also be noted that very occasionally the second derivative test fails. For an example of this consider the function $y = x^4$. It is a bowl shaped curve with a minimum centred on the origin. Substituting $x = 0$ into the second derivative expression we have

$$\frac{d^2y}{dx^2}\bigg|_{x=0} = 0$$

That is, the curve has a second derivative of zero at its minimum turning point when a positive value would normally be expected.

Let us consider the function $y = x^3 - 7x^2 + 15x$ and find the location of its turning points and their nature. Taking the first derivative and setting it to zero we have

$$\frac{dy}{dx} = 3x^2 - 14x + 15 = 0$$

Solving for x we have

$$(3x - 5)(x - 3) = 0$$

$$x = \frac{5}{3} \text{ or } x = 3$$

i.e. stationary points exist at these x-values. Now taking the second derivative we have

$$\frac{d^2y}{dx^2} = 6x - 14$$

$$\therefore \frac{d^2y}{dx^2}\bigg|_{x=\frac{5}{3}} = 6\left(\frac{5}{3}\right) - 14 = -4$$

$$\frac{d^2y}{dx^2}\bigg|_{x=3} = 6(3) - 14 = 4$$

Thus, there is a maximum turning point located at $x = \frac{5}{3}$ and a minimum at $x = 3$.

A turning point on a graph is not necessarily the extremum. There may exist end points of the curve that are more extreme than the turning points. In such instances the turning points are referred to as a *local* maximum or minimum, as the case may be. Whereas the greatest value is called the *global maximum*, whether it is an end point or a turning point. The least value is referred to as the *global minimum*. We are ignoring here curves stretching to infinity and considering only those graphs that have end points.

It can be helpful to find not only the turning points but also any *points of inflection*. These points are locations on a curve where a concave down meets a concave up. For these points the rate of change of the gradient is zero. That is, the second derivative equals zero at points of inflection. Also, it may turn out that a stationary point is neither a maximum nor a minimum but a point of inflection. These points are known as *stationary points of inflection*. That is, when the point of inflection has a zero gradient.

To locate any points of inflection, simply find the second derivative and equate to zero. For example, taking the second derivative of the previous function and equating to zero gives

$$\frac{d^2y}{dx^2} = 6x - 14 = 0$$

$$\therefore x = \tfrac{7}{3}$$

Substituting for x into the original function locates the point of inflection at $\left(\tfrac{7}{3}, 9\tfrac{16}{27}\right)$.

It may also be useful to determine where a curve is increasing or decreasing. When a curve is constantly increasing then the slope, or gradient, is positive. For a curve constantly decreasing the gradient is negative. Regions on the graph where the gradient is positive or negative can be found by setting

$$\frac{dy}{dx} > 0 \ \text{ or } \ \frac{dy}{dx} < 0$$

and finding the interval for which this holds.

Other points convenient for graphing curves are the axis intercepts. Since an intercept necessarily has one ordinate equal to zero then calculating the value of the intercept is generally quite easy. For the x-intercepts let $y = 0$ in the equation and solve for x and vice versa for finding the y-intercept.

More complex graphs may involve vertical and horizontal asymptotes. An asymptote is a line at which the curve will continually approach but never quite reach. Vertical asymptotes, otherwise known as *poles*, will exist at x values which result in a zero on the denominator of the function. For example,

$$y = \frac{3}{(x + 2)(x - 3)}$$

has vertical asymptotes at $x = -2$ and 3. The horizontal asymptotes exist where

$$\lim_{x \to \infty} f(x) = 0$$

Example 7.1. *Find the global and local maxima and minima of the function $y = x^3 - 12x$ over the domain [-3, 5]. Also locate the point of inflection.*

Differentiating,

$$\frac{dy}{dx} = 3x^2 - 12 = 0$$

$$\therefore x^2 = 4$$

$$x = \pm 2$$

i.e. stationary points exist at

$$y(\pm 2) = (\pm 2)^3 - 12(\pm 2)$$

$$= \pm 8 \mp 24 = -16 \text{ or } 16$$

i.e. $(2, -16), (-2, 16)$

Determining the nature of the stationary points we differentiate again:

$$\frac{d^2y}{dx^2} = 6x$$

$$\therefore \frac{d^2y}{dx^2}\bigg|_{x=\pm2} = \pm12$$

$$\therefore (2, -16) \text{ is a minimum}$$

$$(-2, 16) \text{ is a maximum T.P.}$$

Testing the end points we have

$$y(-3) = (-3)^3 - 12(-3) = -27 + 36 = 9$$

$$y(5) = 5^3 - 12(5) = 125 - 60 = 65$$

So there exists a local and global minimum at (2, -16), a local maximum at (-2, 16) and a global maximum at (5, 65). Now let

$$\frac{d^2y}{dx^2} = 6x = 0$$

$$\therefore \text{ point of inflection at } x = 0$$

$$\text{i.e. at point } (0, 0)$$

☐

Example 7.2. *With the help of calculus techniques sketch the function* $g(x) = \frac{1}{24}x^4 + \frac{1}{4}x^3 - \frac{43}{24}x^2 + 2.$

Finding turning points:

$$g'(x) = \frac{1}{6}x^3 + \frac{3}{4}x^2 - \frac{43}{12}x = 0$$

$$2x^3 + 9x^2 - 43x = 0$$

$$x(2x^2 + 9x - 43) = 0$$

i.e.

$$\therefore x = 0 \text{ or } x = \frac{-9 \pm \sqrt{9^2 - 4(2)(-43)}}{4}$$

$$= \frac{-9 \pm \sqrt{425}}{4}$$

$$= \frac{-9 \pm 5\sqrt{17}}{4} = 2.9 \text{ or } -7.4$$

Determining y-value of the turning points:

$$g(0) = 2$$

$$g(-7.4) = -72.5$$

$$g(2.9) = -4.02$$

So, turning points exist at $(-7.4, -72.5), (0, 2), (2.9, -4.02)$. Now,

$$g''(x) = \frac{1}{2}x^2 + \frac{3}{2}x - \frac{43}{12}$$

$$g''(-7.4) = \frac{1}{2}(-7.4)^2 + \frac{3}{2}(-7.4) - \frac{43}{12} = 12.7 > 0 \therefore \text{min TP}$$

$$g''(0) = -\frac{43}{12} < 0 \therefore \text{max TP}$$

$$g''(2.9) = \frac{1}{2}(2.9)^2 + \frac{3}{2}(2.9) - \frac{43}{12} = 4.97 > 0 \therefore \text{min TP}$$

x-intercepts:

Let

$$p(x) = \frac{1}{24}x^4 + \frac{1}{4}x^3 - \frac{43}{24}x^2 + 2 = 0$$

$$\therefore x^4 + 6x^3 - 43x^2 + 48 = 0$$

Now, by trial and error we find

$$p(-1) = 0$$

Therefore, by synthetic division:

$$-1 \begin{array}{|ccccc} 1 & 6 & -43 & 0 & 48 \\ \hline 1 & 5 & -48 & 48 & 0 \end{array}$$

$$\therefore p(x) = (x + 1)(x^3 + 5x^2 - 48x + 48) = 0$$

Now, let

$$q(x) = x^3 + 5x^2 - 48x + 48$$

We find,

$$q(4) = 0$$

$$\therefore 4 \begin{array}{|rrrr} 1 & 5 & -48 & 48 \\ \hline 1 & 9 & -12 & 0 \end{array}$$

Therefore,

$$(x + 1)(x - 4)(x^2 + 9x - 12) = 0$$

$$\therefore x = -1, 4, \frac{-9 \pm \sqrt{129}}{2}$$

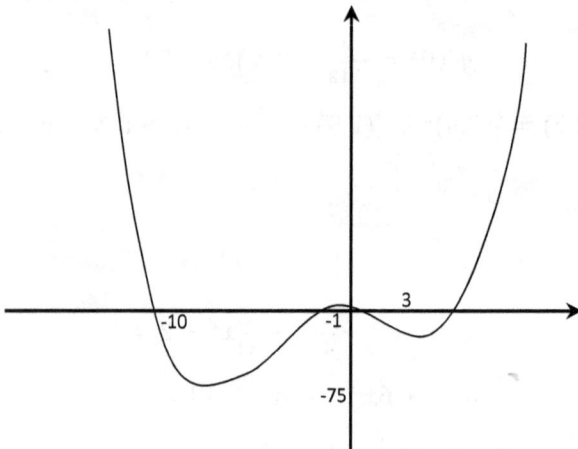

Fig. 7. 2. Showing the graph of a quartic polynomial function.

Plotting these points and drawing a smooth curve connecting them gives an approximate sketch of the polynomial function as in Fig. 7.2. The inflection points were not calculated for this particular graph; it is not necessary to do so though there are some graphs where it may be convenient.

☐

7.2 Optimisation

There are many situations in mathematics, science and industry, where we would like to know the size or location of the optimum value of some quantity. Depending on the situation the required optimum may be either a *maximum* or *minimum*. Differentiation provides us with a powerful tool with which to find these optimum values.

We saw in chapter 1 how the foundations of differentiation had its roots in determining the location of the maximum of a parabolic curve. In addition, we have seen how the minimum turning point may also be located with the benefit of the derivative. If a mathematical function is used to describe some variable quantity, then upon considering its geometrical interpretation, the resulting curve will often involve turning points. By applying the derivative and setting the resulting gradient function to zero, we may find the value and location of these maxima or minima. Thus the optimum value of the variable quantity under consideration may be found.

We can break down an optimisation problem into a series of relatively simple steps:

1. Assign pronumerals to the relevant quantities. Determine an equation for the quantity to be optimised in terms of a single independent variable. E.g. finding a function y in terms of x. A diagram may prove useful here.

2. Take the derivative with respect to the independent variable.

3. Let the derivative equal zero and solve the resulting equation. E.g. solve for x.

4. Test whether there exists a maximum or minimum at these x values.

5. Substitute the relevant x-value into the original equation to determine the optimum value.

Note that the domain and range of the quantities in question need to be considered, as an extrema may lie outside allowable values.

As is often the case, the actual calculus behind an optimisation problem may well be the easiest part. Finding the equation and performing any required algebra may prove the most difficult. Strategies, such as drawing a diagram and assigning pronumerals to unknown quantities as a first step, may prove useful in cases where the optimisation equation is difficult to find.

Example 7.3. *A market gardener has 50 m of fencing material and wishes to fence off an area of soil in the shape of a circle sector. What is the radius and angle of the sector if the gardener would like to maximise its area?*

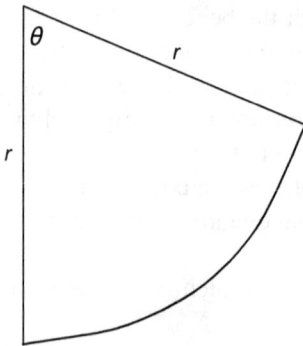

The area of a sector formula gives,

$$A = \frac{\theta°}{360} \cdot \pi r^2$$

with,

$$2r + \frac{\theta}{360} \cdot 2\pi r = 50$$

i.e.

$$\frac{\theta}{360} \cdot \pi r = 25 - r$$

i.e.

$$\theta = \frac{360(25 - r)}{\pi r}$$

Substituting into the area function gives

$$A = \frac{360(25 - r)}{\pi r} \cdot \frac{\pi r^2}{360} = (25 - r)r$$

Differentiating with respect to r gives

$$\frac{dA}{dr} = 25 - 2r$$

Setting to zero we have,

$$25 - 2r = 0$$

$$-2r = -25$$

$$r = \frac{25}{2}$$

Hence we have shown that there exists an extremum at $r = 25/2$. The second derivative is negative and therefore we know the area is maximum at this radius. Substituting this radius value into our determined function for the sector angle, the optimum angle size is therefore given by

$$\theta = \frac{360(25 - \frac{25}{2})}{\pi \cdot \frac{25}{2}} = \frac{360}{\pi} \approx 115°$$

□

Example 7.4. *Find the minimum value of the expression* xy^2 *if* x *and* y *are constrained by the equation* $x - y = 2$.

Let

$$z = xy^2$$

Since

$$x - y = 2$$

then

$$y = x - 2$$

and therefore,

$$z = x(x - 2)^2$$

The minimum value requires the first derivative to be zero. So,

$$\frac{dz}{dx} = x \cdot 2(x - 2) + (x - 2)^2 = 0$$

i.e.

$$(x - 2)(2x + x - 2) = 0$$

$$(x - 2)(3x - 2) = 0$$

$$\therefore x = 2 \text{ or } \frac{2}{3}$$

Now, testing the nature of the critical values:

$$\frac{d^2z}{dx^2} = (x - 2) \cdot 3 + (3x - 2) \cdot 1 = 6x - 8$$

So,

$$\left.\frac{d^2z}{dx^2}\right|_{x=2} = 6(2) - 8 = 4 \quad \therefore \text{minimum}$$

That is, xy^2 has a minimum value of $2(2 - 2)^2 = 0$.

□

Example 7.5. *A long strip of sheet metal of width 40 cm is to be bent to form an open gutter of depth 10 cm and with a symmetrical trapezoidal cross-section. What are the dimensions of the trapezium to ensure maximum flow of rainwater?*

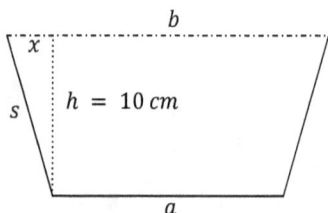

We are required to maximise volume. The volume of a trapezoidal prism is given by

$$V = \frac{1}{2}(a + b)hl$$

where l is the length of the gutter. However, the flow rate will only be dependent on the cross-sectional area and thus we only need to maximise area $A = \frac{1}{2}(a + b)h$. From the diagram we see that

$$a + 2s = 40$$

and

$$b = a + 2x$$

Therefore the area may be expressed as

$$A = \frac{1}{2}(a + a + 2x) \cdot 10$$

$$= 10(40 - 2s + x)$$

From Pythagoras's theorem we find the relationship

$$s = \sqrt{x^2 + 100}$$

Hence,

$$A = 10(40 - 2\sqrt{x^2 + 100} + x$$

Differentiating with respect to x and equating to zero we have

$$\frac{dA}{dt} = 10\left[-(x^2 + 100)^{-\frac{1}{2}} \cdot 2x + 1\right] = 0$$

$$\therefore (x^2 + 100)^{-\frac{1}{2}} \cdot 2x = 1$$

$$2x = (x^2 + 100)^{\frac{1}{2}}$$

$$4x^2 = x^2 + 100$$

$$3x^2 = 100$$

$$x = \pm\frac{10}{\sqrt{3}}$$

The negative solution can be ignored since that would necessitate $a > b$. So,

$$s = \sqrt{\frac{100}{3} + 100} = \sqrt{\frac{400}{3}} = \frac{20}{\sqrt{3}}$$

Therefore,

$$a = 40 - 2 \cdot \frac{20}{\sqrt{3}} = \frac{40\sqrt{3} - 40}{\sqrt{3}} = \frac{40(\sqrt{3} - 1)}{\sqrt{3}} \text{ cm}$$

and

$$b = \frac{40(\sqrt{3} - 1)}{\sqrt{3}} + 2 \cdot \frac{10}{\sqrt{3}} = \frac{40\sqrt{3} - 20}{\sqrt{3}} = \frac{20(2\sqrt{3} - 1)}{\sqrt{3}} \text{ cm}$$

□

7.3 Related Rates

There are certain situations where two *related* quantities vary relative to a third quantity, such as time. For example, the radius of a cylindrical object may be changing over time and it is desired to find the *rate* at which the volume is changing at some specific radius.

Solving *related rates* problems is normally a fairly formulaic procedure that can be done in a series of relatively simple steps as follows:

1. Assign pronumerals to the unknown quantities.

2. Draw a labelled diagram—a diagram is usually very useful to help understand the situation.

3. Find an equation relating the principal unknown (e.g. y) with another unknown quantity (e.g. x) and take the derivative.

4. Find the derivative relative to a third unknown, usually time, utilising the *chain rule*. E.g. $\frac{dy}{dt} = \frac{dy}{dx} \cdot \frac{dx}{dt}$.

5. Evaluate this derivative at the required point.

How to implement these steps is best illustrated by example. Consider the following worked exercises:

Example 7.6. *The radius of a circle is increasing at a rate of 2 cm/s. At what rate is the circumference and area of the circle increasing when the radius equals 20 cm?*

Let r, C, A and t represent the radius, circumference, area and time respectively, with $\frac{dr}{dt} = 2$. Now,

$$C = 2\pi r$$

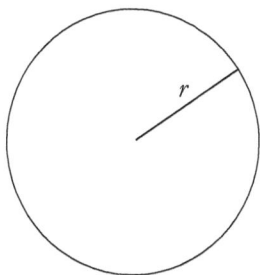

Differentiate C with respect to r:

$$\frac{dC}{dr} = 2\pi$$

Differentiate C with respect to t:

$$\frac{dC}{dt} = \frac{dC}{dr} \cdot \frac{dr}{dt}$$

$$= 2\pi \cdot 2 = 4\pi \text{ cm/s}$$

Also,

$$A = \pi r^2$$

Therefore,

$$\frac{dA}{dr} = 2\pi r$$

and

$$\frac{dA}{dt} = \frac{dA}{dr} \cdot \frac{dr}{dt} = 2\pi r \cdot 2 = 4\pi r$$

Therefore at $r = 20$ cm the rate of change of the area of the circle is 80π cm^2/s.

$\square$

Example 7.7. *A liquid industrial chemical is being pumped at a rate of 100 litres per second into an inverted cone shaped tank of depth 6 m with a maximum radius of 3 m. At what rate is the depth of the liquid increasing (i) at a depth of 2 m? (ii) at a volume of 10π m^3?*

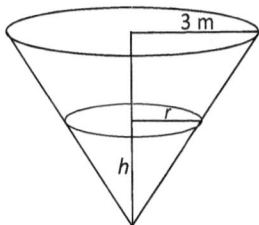

$$\frac{dV}{dt} = 100 \text{ L/s} = 100 \cdot \frac{\frac{1}{1000} \text{ m}^3}{\text{s}}$$

$$= 0.1 \text{ m}^3/\text{s}$$

From similar triangles we have

$$\frac{h}{r} = \frac{6}{3} = 2$$

i.e.

$$r = \tfrac{1}{2}h$$

Now, the volume is given by

$$V = \tfrac{1}{3}\pi r^2 h$$

Substituting for r we have

$$V = \tfrac{1}{3}\pi \left(\tfrac{1}{2}h\right)^2 h = \tfrac{1}{3}\pi \cdot \tfrac{1}{4}h^2 \cdot h = \tfrac{1}{12}\pi h^3$$

Differentiating with respect to h

$$\frac{dV}{dh} = \tfrac{1}{4}\pi h^2$$

Using the chain rule we have

$$\frac{dh}{dt} = \frac{dh}{dV} \cdot \frac{dV}{dt}$$

$$= \frac{1}{\tfrac{1}{4}\pi h^2} \cdot 0.1$$

$$= \frac{4}{\pi h^2} \cdot \frac{1}{10} = \frac{2}{5\pi h^2}$$

So,

$$\left.\frac{dh}{dt}\right|_{h=2} = \frac{2}{5\pi \cdot 2^2} = \frac{1}{10\pi}\, \text{m/s} \approx 3.2 \text{ cm/s}$$

For the second part of the question, since

$$V = \tfrac{1}{12}\pi h^3$$

then

$$\tfrac{1}{12}\pi h^3 = 10\pi$$

$$\therefore h^3 = 120$$

$$h = \sqrt[3]{120} = 2\sqrt[3]{15}$$

Now,

$$\left.\frac{dh}{dt}\right|_{h=2\sqrt[3]{15}} = \frac{2}{5\pi\left(2\sqrt[3]{15}\right)^2} = \frac{1}{20\pi \cdot 15^{\frac{2}{3}}} \approx 2.6 \text{ mm/s}$$

□

Example 7.8. *A man 2 m in height is walking at night in a straight line at a speed of 1 m/s away from a street light that is 4.2 m above the ground. At what rate is his shadow lengthening when he reaches a point 10 m distance from the light assuming he is walking on flat ground?*

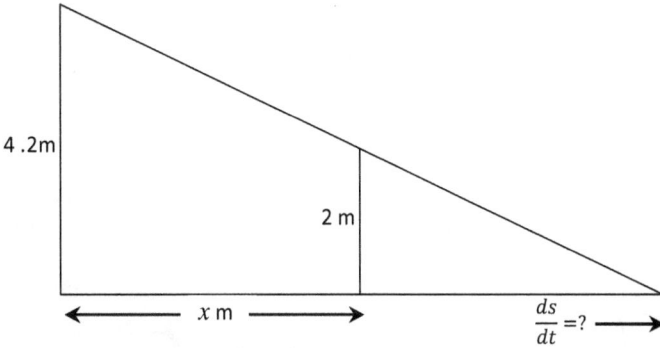

Let s represent the length of the man's shadow and x his distance from the base of the light. From the diagram above, similar triangles give us

$$\frac{x + s}{4.2} = \frac{s}{2}$$

$$\therefore x + s = 2.1s$$

$$1.1s = x$$

$$s = \frac{10}{11}x$$

Now,

$$\frac{ds}{dt} = \frac{ds}{dx} \cdot \frac{dx}{dt} = \frac{10}{11} \cdot 1 = 0.9\overline{09}$$

i.e. the shadow is lengthening at a rate of approximately 0.91 m/s and is independent of the man's distance from the street light.

□

Exercise 7

1. For what domains are the following functions (i) decreasing, (ii) increasing?

   a) $f(x) = x^2 - 4x + 3$
   b) $g(x) = x - 2x^2$
   c) $A(x) = 2x^2 - x^3 + 8x$
   d) $B(x) = x^4 + 8x$

2. For which intervals is the function $\theta(t) = \cos t + \sin t$ decreasing for the domain $\left[-\frac{\pi}{2}, 2\pi\right]$?

3. Is the function $H(w) = \left(x + \frac{1}{x}\right)^2$ increasing, decreasing or neither at the point (-1, 8)?

4. Find the intervals where the function $g(x) = \frac{3-x}{2x^2}$ is increasing and where it is decreasing. Sketch the curve with the help of calculus techniques.

5. Show that for the function $f(x) = x^5 - 10x^4 + \frac{80}{3}x^3 - 9$ it is increasing everywhere.

6. Find where the following functions are increasing and where they are decreasing:

   i) $y = \frac{x^2-5x}{(x+2)^2}$ ii) $y = 3x - \frac{5}{x}$

7. Graph the following functions without the use of a computer based graphing program:

   a) $f(x) = x^2 + 4x - 5$
   b) $g(x) = 9x - 3x^2 - 6$
   c) $h(x) = x^3 - 4x$
   d) $A(x) = x^3 - 2x^2 - 19x + 20$
   e) $E(\theta) = \sin^2 \theta + \cos \theta$; $[0,2\pi]$

8. Find the extrema of the following functions over the given domains.

a) $f(x) = 9 - x^2$;   [-2, 1]

b) $g(x) = x^2 - 5x + 4$;   [-1, 6]

c) $h(x) = (x - 3)^3 + 2$;   [0, 5]

d) $G(x) = \frac{1}{2}(x^3 - 10x)$;   [-5, 3]

9. Find any local maxima and minima and points of inflection of the functions below. Find any other helpful points and draw a sketch of the associated curves and clearly indicate any asymptotes.

a) $g(x) = x^3 - 10x + 4$

b) $y = 2(x - 3)^4$

c) $\theta(x) = \sqrt{x} - \frac{1}{\sqrt{x}}$, $x > 0$

d) $f(x) = x^2 + \frac{2}{x^2}$

e) $H(x) = \left(\frac{(x-2)(x+3)}{x-4}\right)^2$

f) $y = (x^2 + x - 3)^2$

g) $f(x) = \frac{5(x-2)^2}{(x+2)^3}$

h) $g(x) = \frac{2x}{(x+6)(x-4)}$

i) $h(x) = \frac{(2-x)^3}{x+3}$

10. An experimental electromagnetic rail gun fires a projectile vertically upwards to a height y m after t s, given by the equation

$$y = 149t - 4.9t^2$$

To what height does the projectile rise assuming no air resistance?

11. The fuel economy of the average car is approximated by the parabolic function

$$E = -0.0012v^2 + 0.21v + 2.2$$

where E is in km/L and the speed v is in km/h. Find the most fuel efficient speed. What is the fuel economy at this speed? Express your answer in litres per 100 km.

12. Find two real numbers whose sum is 12 and whose product is a maximum.

13. Which point along the line $2y - x = 4$ is the closest to the origin?

14. What are the dimensions of a rectangular area that is required to be maximised given its perimeter is 500 m ?

15. A rectangular shaped pool of area 60 square metres is to be paved all around to a width of 2 metres. What is the required dimensions of the pool such that the total area of pool and paving is kept to a minimum?

16. An open box is to be made from a sheet of metal by cutting equal sized squares from each corner and folding the resulting tabs to create vertical sides. The dimensions of the sheet metal are 75 cm × 50 cm. If a box of maximum volume is required, what is the size of the squares cut from the metal?

17. A food company is required to produce tin cans of volume 375 ml. What are the dimensions of the can that minimises the amount of tin used?

18. A rectangular greenhouse is to be built on the side of a shed using the shed wall as one side of the greenhouse. If 30 metres of material is available with which to construct the remaining walls of the greenhouse, what dimensions would maximise the area?

19. Some polynomial functions continually increase in value and thereby their gradient, though changing in value, will remain positive. Find the minimum gradient of the polynomial function $y = x^3 - 4x^2 + 10x + 5$.

20. At 8 am a ship is 75 km due west of another ship. The first ship is sailing at a speed of 25 km/h in the easterly direction while the second ship is sailing southwest at a rate of 20 km/h. At what time will the two ships be at their closest and how far apart are they at this time?

21. A cylindrical telescope, open at one end, is constructed from material of area 1250π cm^2. Find the greatest volume of the telescope.

22. Find the maximum possible area of a rectangle that has sides lying on the Cartesian axes and a vertex coincident with the curve

$$y = \sqrt{9 - x^2}$$

23. What is the largest volume of a cylinder that just fits into a cone of 30 cm height and base diameter 28 cm.

24. A closed rectangular box is to have a volume of 25 L. The base of the box needs to be the strongest section and is therefore made from a material that is 50% more expensive than the top. The sides need to be relatively strong but can be made from material that is ¾ the cost of the bottom. Find the dimensions of the box that will minimise cost given that its width equals to its depth.

25. An open water channel with an isosceles triangle cross-section is to be constructed. The sides are required to be 20 cm. What angle between the sides will maximise the flow of water?

26. An underground mine requires a 5 cm diameter pipe to be moved along a shaft 1 m wide that turns at right angles into another shaft of width 75 cm. Determine the longest piece of pipe that can just fit around the corner. Assume the pipe remains horizontal.

27. The height of an isosceles triangle of base 2 cm is increasing at a rate of $2\sqrt{2}$ cm/s. At what rate are the congruent sides lengthening once the height reaches 10 cm?

28. As air is pumped into a spherical balloon its radius increases at the rate of 0.5 cm/s. At what rate is the surface area and volume increasing when the balloon's diameter reaches 20 cm?

29. Out in the country a car travelling at 110 km/h south passes through an intersection at the moment another car on the intersecting road is travelling at 80 km/h east. At this point in

time the second vehicle was at a distance of 80 m. At what rate are the two vehicles moving apart after 1 min?

30. A fisherman standing on a jetty 3 m above the water is reeling a fish in. The end of the fishing rod is 1 m above the fisherman's head and the line is being reeled in at 1.5 m/s. At what speed is the fish being hauled in at when it is 10 m from the base of the jetty? Assume the fish stays near the surface of the water.

31. Sand is pouring from a chute at an industrial facility at a rate of 0.9 cubic metres per second, forming a conical pile. The angle of repose (the angle between the base and side of a cone) of the sand pile is given by $\tan \theta = \frac{1}{\sqrt{3}}$. How fast is the pile of sand growing once it has reached 2 m in height?

32. A parabolic shaped bowl is being filled with water at a rate of 10 mL/s. The volume of the bowl is given by $\frac{1}{2}\pi y^2$ where y is the depth in cm. How quickly is the depth of water increasing once the depth reaches 3 cm?

33. Consider the Earth in its orbit about the Sun as shown in the diagram below. The Earth moves in its orbit at 30 km/s. At what rate does the Earth's distance from a fixed point X in space change after revolving through $2\pi/3$ radians from this point?

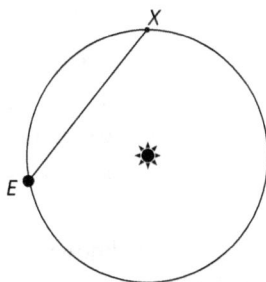

34. A painter stands near the top of a 6.5 m ladder that is leaning against a wall. The ladder starts slipping with the bottom moving away from the base of the wall at 1.75 m/s. At what speed is the painter falling when the top part of the ladder is 0.5 m above the ground.

35. In an investigation into shadow movements for night matches of a ball game, a ball is dropped from the same height as one of the 20 m tall stadium lights and at a horizontal distance of 15 m. The ball falls vertically with its displacement given by $y = 4.9t^2$. With what speed is the shadow of the ball on the ground moving when it is 2.5 m above the ground?

36. A trapezoidal water trough of 0.5 m across at the bottom, 0.95 m at the top and 0.75 m deep, is 5 m long and being filled at 1.2 m³/s from a hose. How quickly is the water level rising when the depth reaches 50 cm?

37. A spherical shaped hailstone 0.8 cm across is blown back up high into the atmosphere by strong draughts. At the colder upper atmosphere temperatures ice crystallisation starts to coat the hailstone with more layers of ice at a growth rate of 10 mm³/s. At what rate is the thickness of these new ice layers growing when they have become 2 mm thick? How fast is the surface area of the hailstone increasing when it reaches 1.95 cm in size?

38. Water is being added to an inverted cone shaped tank, vertex angle 60° at a rate of 1 L / 5 min while concurrently water is leaking at a rate of 10 cm³/s. Find how fast the water level is dropping when the water depth is 30 cm.

39. A laser beam pointed towards the Moon moves in an arc of a circle at a rate of π rad/s. The distance from the laser to the closest point X on the Moon is 3.844×10^5 km. How fast is the spot of light projected onto the Moon's surface moving at (i) 0 km and (ii) 1000 km from point X given the Moon's radius is 1737 km?

40. The Gaussian curve, or what is known as the normal curve in statistics, is described by the function

$$y = \frac{1}{\sqrt{2\pi}} e^{-x^2}$$

Find the coordinates of the points of inflection and the maximum.

41. A cubic curve with the general polynomial function $y = ax^3 + bx^2 + cx + d$ has turning points at $(-2, 5)$ and $(4, -6)$. The point of inflection lies at $(1, 1)$. Find the polynomial.

42. Find the angles for the extrema of the function

$$y = \frac{3}{4}\tan^4 \theta - \frac{8}{3}\tan^3 \theta + \frac{3}{2}\tan^2 \theta + 2\tan \theta - 1$$

in the interval $-\frac{\pi}{4} \le \theta \le \frac{\pi}{4}$.

43. What is the closest and furthest point on the circle $x^2 + y^2 = 9$ to the point $(1, 1)$?

44. The sum of the radius and height of a cone is 7. Assigning ϕ to the semi-vertical angle, find $\tan \phi$ if it required the surface area of the cone be a maximum.

45. The length of the hypotenuse of the triangle formed from the Cartesian axes and a straight line is $2\sqrt{5}$. Find the equations of the lines that give a maximum perimeter of the triangle.

46. Show that the volume of the largest cone that can be placed inside a sphere is in ratio 8:27 to its volume.

47. A particle moving along the x-axis has its displacement given by $x = \alpha t^3 + \beta t^2 - 10$, where t represents time and α, β are constants. Find these constants knowing that the maximum velocity of 57 is achieved when $t = 3$.

48. A rocket made from an open cylinder and a cone placed at the end is to be made from 200π cm^2 aluminium sheeting. If the height of the cylinder section is required to be 30 cm, by using a computer to aid calculation, find the radius and height of the cone section if the volume of the rocket is to be maximised.

49. The relation $x^3 + y^3 = 3xy$ describes a curve known as the Folium of Descartes. As the name implies it was discovered by Descartes, with 'folium' being Latin for 'leaf', since the curve forms a loop in the first quadrant of the Cartesian plane. Find the x-ordinate of the maximum height of the folium part of the curve.

50. Sketch the parabola $y = x^2 - 4$ and draw the triangle formed by a horizontal chord of the curve and the point (0, 6). Now, determine the maximum area of the triangle.

51. A manufacturing company determines that an industrial part it manufactures is sold at a wholesale price that varies according to the number of units sold, x, as $\$(2 - 5 \times 10^{-3}x)$ per unit. The total cost of production is given by $\$(5 \times 10^{-5}(x - 100)^2 + 0.5)$. Use this information to find the production run that maximises profit.

52. A triangle is inscribed into a circle of radius r and centre O with the triangle's vertices A, B, C coincident with the circumference. Assigning the size of angle AOB as 2θ show the area of the triangle is given by $r^2 \sin\theta(\cos\theta + 1)$. From this expression find the maximum area of the triangle in terms of r.

53. Find the maximum perimeter of a trapezium inscribed in a semi-circle of radius 4 cm.

54. A body in rectilinear motion has its displacement from the origin given by

$$x = t^3 - 12t^2 + 36t$$

Determine
   i.   When the velocity is instantaneously equal zero.
   ii. The displacement at these times.
   iii. The time at which acceleration is instantaneously zero.
   iv. The times for which velocity is increasing.
   v.   The distance traversed between the zero velocity times.

55. Find the minimum area of triangle PQR formed by the point $P(x, y)$ on line $2x + 3y = 2$, and points $Q(-1, 0)$ and $R(0, -0.5)$.

56. The shape of the exhaust plume of a rocket in the Earth's lower atmosphere can be modelled approximately by a cone of vertex angle $\pi/3$. An experimental reusable rocket is descending vertically towards the ground at a speed of 25 m/s.

At what rate is the portion of ground in contact with the plume decreasing at the instant the base of the rocket is 100 m from the landing pad.

57. Some functions exist with turning points but no inflection points.

(i) Find the turning points for

$$f(x) = x^2 - 5 + \frac{16}{x^2}$$

and show there are no points of inflection.

(ii) For functions of the form

$$f(x) = ax^2 + b + \frac{c}{x^2}$$

a, b, c constants, what requirement is there to guarantee no inflection points?

Answers

1a) i) $x > 2$ ii) $x < 2$ b) i) $x < \frac{1}{4}$ ii) $x > \frac{1}{4}$ c) i) $\frac{2}{3}(1-\sqrt{7})<x<\frac{2}{3}(1+\sqrt{7})$
ii) $x < \frac{2}{3}(1 - \sqrt{7}) \cup x > \frac{2}{3}(1 + \sqrt{7})$ d) i) $x > -\sqrt[3]{2}$ ii) $x < -\sqrt[3]{2}$
2) $\frac{\pi}{4}<t<\frac{5\pi}{4}$
3) neither
4) decreasing for $x < 6$, increasing for $x > 6$
5) all reals
6) i) decreasing $x < \frac{10}{9}$, increasing $x > \frac{10}{9}$ ii) increasing all reals, $x \neq 0$
8) a) local and global maximum at $(0, 9)$; global minimum at $(-2, 5)$
b) local and global min at $(5/2, -2\frac{1}{4})$; global max at $(-1, 10), (6, 10)$
c) global min at $(0, -25)$; global min $(5, 10)$ d) local max
$\left(-\sqrt{\frac{10}{3}}, -\frac{10\sqrt{30}}{9}\right)$, local min $\left(\sqrt{\frac{10}{3}}, \frac{20\sqrt{30}}{9}\right)$; global max $(3, -3/2)$, global
min $(-5, -75/2)$
9) a) local min $(1.83, -8.17)$; local max $(-1.83, 16.2)$; point of inflection $(0, 4)$ b) local min $(3, 0)$ c) no stationary points nor point of inflection   d) min TPs $\left(-\sqrt[4]{2}, 2\sqrt{2}\right), \left(\sqrt[4]{2}, 2\sqrt{2}\right)$ e) max $(2, 0), (-3, 0)$, $(4 - \sqrt{14}, 2.3)$; min $(4 + \sqrt{14}, 271.7)$; inflection pts $(-10.8, 45.5)$, $(-1.24, 0.6)$ f) min $\left(\frac{-1\pm\sqrt{13}}{2}, 0\right)$; max $(-1/2, 10\ 9/16)$; inflection pts $(-1.54, 4.7), (0.54, 4.7)$ g) min $(2, 0)$; max $(-26/7, -29.25)$; in-

flections points $(3.8, 0.3), (-4.4, -14.8)$ h) min $(-2\sqrt{6}, 1), (2\sqrt{6}, 1)$;
inflection point $(-0.66, 0.05)$ i) min $(-1\,\%, 42\,3/16)$; stationary pt
of inf. $(2, 0)$

10) 1.33 km

11) 8.8 L/100 km

12) $x = 6, y = 6$

13) $(-4/5, 1\,3/5)$

14) 125 m × 125 m

15) $2\sqrt{15}$ m × $2\sqrt{15}$ m

16) 12.5 cm

17) $h = 5\sqrt[3]{\frac{12}{\pi}}$ cm, $r = \sqrt[3]{\frac{750}{4\pi}}$ cm

18) 15 m × 7.5 m   19) 4 2/3

20) 8:47 am, 45.4 km

21) $\pi\left(\frac{1250}{3}\right)^{3/2}$ cm^3

22) 4.5   23) $\frac{7840}{9}\pi$ cm^3

24) 33.02 cm   25) 90°   26) 2.37 m

27) 2.81 cm/s

28) 40π cm^2/s, 200π cm^3/s

29) 91.4 km/h   30) 1.4 m/s

31) $\frac{9}{80\pi}$ m/s

32) $\frac{10}{3\pi}$ cm/s

33) 15 km/s

34) 13 m/s

35) 18.1 m/s   36) 30 cm/s

37) $\frac{5}{72\pi}$ mm/s, 2.05 mm^2/s

38) 0.067 mm/s

39) i) 387874π km/s ii) 1.45×10^6 km/s

40) max $\left(0, \frac{1}{2\pi}\right)$, pts of inflection $\left(\pm\frac{1}{\sqrt{2}}, \frac{1}{\sqrt{2\pi e}}\right)$

41) $y = \frac{2}{27}x^3 - \frac{2}{9}x^2 - 48x + \frac{1319}{27}$ 42) $\frac{\pi}{4}, -0.32, 1.11$

43) $\left(\frac{3}{\sqrt{2}}, \frac{3}{\sqrt{2}}\right), \left(-\frac{3}{\sqrt{2}}, -\frac{3}{\sqrt{2}}\right)$

44) 2   45) $y \pm x = \pm\sqrt{10}$

46) 8/27   47) -19/9, 19

48) 3.16 cm, 0.92 cm

49) $\sqrt[3]{2}$ 50) $\frac{20\sqrt{30}}{9}$ 51) 200   52) $\frac{3\sqrt{3}}{4}r^2$

53) 20 cm

54) i) 2, 6 ii) 32, 0 iii) 4 iv) $t > 4$ v) 32

55) 0.551   56) -5236 m^2/s

57) i) $(\pm 2, 3)$ ii) $c < 0$

Index

A

acceleration, 30
Apollonius, 3, 13
asymptote, 105

C

calculus, 14
chain rule, 53, 60
composite function, 59
critical point. *See* stationary point
curve sketching, 101

D

Derivative, 1, 14, 21
  exponential functions, 92
  inverse trigonometric
    functions, 84
  logarithmic function, 93
Descartes, 1, 14
differentials, 37
differentiation, 1, 17, 29, 47
  - trigonometric functions, 68
displacement, 33

E

Euler's number, 92

F

Fermat, 1, 10, 14
function, 79

G

gradient, 2, 14, 36

I

implicit differentiation, 79
increments, 37

K

Kepler, 14

L

Lagrange, 18
Leibniz, 1, 16, 47
line of symmetry, 12
logarithmic differentiation, 96

M

maxima and minima. *See*
  maximum and minimum
maximum, 109
  global, 104
  local, 104
maximum and minimum, 10
minimum, 109
  global, 104
  local, 104

N

natural logarithm, 93
Newton, 14, 47

normal, 36

O

optimisation, 109

P

parabola, 2, 11, *See*
parametric equations, 62
Pascal, 19
percentage error, 39
point of inflection, 104
poles. *See* asymptotes
product rule, 47

Q

quotient rule, 47, 57

R

rates of change, 30
rectilinear motion, 29
related rates, 114
relation, 79

S

second derivative test, 101
sign test, 101
stationary point, 101

T

tangent, 2, 10, 12, 34
turning point, 10

V

velocity, 29